*In the Spirit of the Earth*

# In the Spirit of the Earth

~~~~~~~~~~

## Rethinking History and Time

Calvin Luther Martin

The Johns Hopkins University Press
Baltimore and London

This book has been brought to publication with the generous assistance of the Chester Kerr Publication Fund.

The Johns Hopkins University Press
701 West 40th Street
Baltimore, Maryland 21211-2190
The Johns Hopkins Press Ltd., London

Acknowledgments of permission to reprint previously published
works are found on page 159.

The paper used in this book meets the minimum requirements
of the American National Standard for Information Sciences
—Permanence of Paper for Printed Library Materials,
ANSI Z39.48-1984.

Library of Congress Cataloging-in-Publication Data
Martin, Calvin.
In the spirit of the earth : rethinking history and time /
Calvin Luther Martin.
p.    cm.
Includes bibliographical references and index.
ISBN 0-8018-4358-8 (alk. paper)
1. Hunting and gathering societies.   2. Human evolution—
Philosophy.   3. Human ecology—Philosophy.   I. Title.
GN388.M37   1992
304.2'01—dc20      91-38384

*Nina*

# Contents

∧∧∧∧∧∧∧∧

# Acknowledgments

∿∿∿∿∿∿

I have accumulated large intellectual debts. Loren Eiseley has been much on my mind. Someday, I hope, scholars will place him among the towering figures in Western thought. I was reading him in Princeton while he was dying in Philadelphia, and I never met him. On the other hand, he was something of a recluse, and probably would have discouraged a meeting, especially then. Thoreau, too, has been in my thoughts—another solitary, and Eiseley's spiritual mentor. Over the years I have profited as well from reading C. S. Lewis, the Kiowa novelist and philosopher N. Scott Momaday, Joe Meeker's columns in *Minding the Earth Quarterly,* Paul Shepard's books on the themes covered in this one, Frederick Turner (the contemporary historian), Mircea Eliade, Marshall Sahlins, several formidable French thinkers —Michel Foucault, Marcel Detienne, Claude Lévi-Strauss, Lucien Lévy-Bruhl—and the poetry and prose of Gary Snyder and Robert Bringhurst. If anyone is going to reinvent humanity, surely it will be the poets.

I have been lecturing on these themes for the past seventeen years and am grateful to my Rutgers students for helping me refine my ideas. My students have become good friends and stimulating colleagues. Various scholars in and out of Rutgers have read the manuscript and contributed valuable thoughts: Wendy Ashmore, Carmel Schrire, Philip J. Greven, Thomas Slaughter, Paul G. E. Clemens, Kevin Dann, and Joseph Moore. Scott Kushner and Alain Hénon deserve special acknowledgment. Barbara Lamb (my editor), Jack Goellner (director of the Johns Hopkins University Press), and Henry Y. K. Tom (executive editor) have been enormously stimulating and encouraging.

Nina Pierpont's Ph.D. in population biology and ecology and her M.D. degree have been put to good use herein, along with her willingness to listen to these chapters read time and again. Spouses are kind. So, too, have been the National Endowment for the Humanities, the John Simon Guggenheim Foundation, and the American Council of Learned Societies, all of which have awarded me fellowships over the past decade, fellowships that have assisted materially in the preparation of this book.

Many thanks to all of these individuals and institutions.

*In the Spirit of the Earth*

# 1

## "An inhuman company of elder things"

∿∿∿∿∿∿

### Once We Were Shape-Shifters

Words. I have grown suspicious of them. The irony is that I am paid
handsomely to use them. And use them I do, mostly in delivering
windy lectures to hundreds of university students every year, trying
to convey an understanding of the history of the North American
continent both before and after the European arrival. I have been
holding forth on the subject for years and am growing increasingly
distrustful of what I myself have been saying. Partly because I use the
language of the Anglo-American; they are the words of a Western-
trained, Western-oriented intellect as it seeks to wrap itself around
this place, this landscape, and to convey what has transpired here in
the affairs of humankind. Yet all the while I am mocked by the
knowledge that my words are not the words of the people whom my
European ancestors encountered on this continent. Those earlier
residents pronounced the place, and described the affairs of people,
very differently from the way I do now or my Maryland forebears did
centuries ago. Our differing sets of words have yielded profoundly
variant stories. One of them I flatter myself I know well. The other,
the aboriginal, is the problem. The people native to North America

spoke (and many still speak) of a world, a place, and a way of living
and being, all alien to my Western cast of mind. And yet both stories
seek to describe and communicate the inherent truths of the same
place and man and woman's role in all of that.

Words reflect the way our minds touch the world about us. In
this sense they are a kind of sonar: they orient us. Among the Navajo,
the *Diné*, femaleness is referred to in terms of certain cardinal points
(south and west), specific colors (blue and yellow), times of day
(daylight and twilight) and life (maturation and old age), seasons
(summer and fall), and a precise, highly complex state of being, the
closest word for it being *active*. Maleness invokes a different, complementary, and seemingly opposed constellation of references. Nava-
jos define themselves through a language that makes connections
between what seem to me curiously arranged categories of perception and knowledge. So, too, are other beings—animal beings, plant
beings—inserted into this wheeling landscape of potentiality, where
the "I" knows how essential it is to "become part of it"—*all* of the
dimensions and categories of place.[1] Thus do these people plug
themselves into the powers of place: through carefully chosen words
and speech.

We, too, are forever being admonished to choose our words carefully. But what do we have to choose from? Scan the dictionary in
your brain, or take down such a volume from the shelf and realize
that our speech is loaded; the words resonate with a distinctive *imagination* of the world—an observation one might of course make for
any culture, past or present. (What is the fundamental imagination
of my language for the earth, I ask myself—afraid of the answer.
Hence my anxiety with words.) "As soon as mankind ceases to 're-
verberate' to the world, the sickness penetrates language." Language
becomes "the victim of illusions produced by words."[2]

Words, says Scott Momaday, are names.[3] Yes, possibly. I like to
think of them more as forces that mold the space around me, into
which I then pour my sense of reality and my energies, after the fash-

ion of the amoeba. As the engineer of space, language does more than establish the terms of discourse between people; it may also do much the same thing between people and the myriad other life forms about us. The idea is aboriginal, not mine, and I readily admit I am perplexed by it. Yet it seems to be there when I try to read between the lines as native peoples so matter-of-factly tell Europeans what went wrong: "With the coming of the whites and Christianity the demons of the bush have been pushed back to the north where there is no Christianity. And the conjuror does not exist any more with us, for there is no need of one. Nor is there need for the drum." Another: "Since prayer has come into our cabins, our former customs are no longer of any service; . . . our dreams and our prophecies are no longer true,—prayer has spoiled everything for us." And a third: "The spirits do not come to help us now. The white men have driven them away." Behold the strange, incantatory powers of a speech that can silence the elemental powers around it. How can anyone who wields such words ever hope to become "a part of it?"[4]

There is one other instrument, of commensurate power, that man and woman have brought to bear on the world: hands, capable of grasping and manipulating in a most intricate, delicate fashion. Here, too, as with our use of words, I have become disturbed by the extravagant artifice of my kind, now, in rendering a relationship with the world. Looking back over past ages of human experience I marvel at all that has been wrought by this ten-fingered primate with two mirror-image hands. More than that, I look for a pattern, a message. Better yet, an ethic. For here, as with language, there was a beginning, in a relationship, a dance, between the place itself and the gifted creature. Did the stupendous dexterity of *Homo sapiens* come from the arboreal handedness of an ancestral primate? The suggestion conjures up the cartoon image of some hairy ancestor gracefully swinging through the canopy. A Promethean Tarzan, perhaps. We chuckle, but a careful look at our simian relatives suggests that the idea has considerable merit.

Whatever the origin—I would say evolutionary origin—of our versatile hands, it is clear that nature designed us to be above all talkers and handlers. No other creature on earth is as extensively neuronally wired to perform these two functions. I refer particularly to the sensory-motor cortex: the region of the brain controlling voluntary bodily function. *Homo sapiens* is unsurpassed in the extent of sensory-motor tissue devoted to firing the tongue, lips, larynx, and fingers. And especially the thumb. Mankind is constitutionally a speaker and a tinkerer—born to chat, as one wit put it. Here, then, is a neuroanatomical baseline for any inquiry into the nature of words and artifice. Those of us who have had children are reminded of the two-year-old monster who talks incessantly and grabs everything within reach. Such is the fundamental nature of the human animal.

It is no accident that youngsters have a stunning capacity to learn languages, a capability that becomes sharply, distressingly curtailed just before puberty. Witness our tortured efforts as adults to learn a second language, whereas a youngster pulls it off with consummate ease. Not to despair, parents, your five-year-old had a huge head start. The child's mind is compulsive about acquiring language. Perhaps there was some sort of aesthetic advantage in this over the long haul of evolutionary sorting. Notice I did not say *survival* advantage. (Whether survival or aesthetics is, or has been, the focal issue for *Homo sapiens* will become clearer as I proceed.) Handling things is likewise a compulsion: a touching child is not naughty but responding to an ancient, insistent, and necessary cognitive message. Children uncover (and, as adults, contribute to) the mystery of the world by handling it, in speech and grasp.

But when does grasping become manipulating? When does it become artifice: the purposeful fashioning of objects? And when do tools differ from art? Or do they ever? All are deeply interesting questions. It is even more provocative to ask what our artifice tells us about our image of the world and our vocation within it. At the start of a new semester I ask my students to visit their local hardware store and inspect the shelves with just this in mind.

And that brings me to the strange archaeology of the human mind. Our capacity to imagine things, to image the world about us: its creatures and plants, its fluid and solid architecture, its heavenly vault, and, all the while, to image ourselves somehow, in some capacity, within all of that which lies tantalizing just beyond our skin. Though not beyond our grasp. *Homo* has always sought to engulf all that surrounds it in word and in hand—or artificial extension of the hand.

Nowhere is it written in stone, however, that we must speak of the world and manually rearrange it as we do now, or ever have done. Over the millennia our ancestors made strategic choices about how they would articulate and manipulate, inspired or driven by their fertile imagination of how the world conducted its business and where the human welfare was situated within that larger commonweal. Today we are the heirs of those words and that artifice, those choices and those images—some of them. And therein lies a powerful and timely story. The telling of it will help us, I believe, better understand the context and reverberations of our present speech and artifice of nature, and hence, better understand our current environmental predicament. Which is that we are producing changes that frighten us both in their scope and in their meaning. We admit to being unnerved by what our hands and words have wrought.

~~~~~~~~~~

The frail old man looked distinctly birdlike. Japasa—"Chickadee," they called him—was dying of heart failure. Doctors could do nothing further, but even so they had advised him to remain in the Fort Nelson hospital. Japasa, however, thought otherwise. As the anthropologist put it, "he needed moose meat, wind, stars, his language, and his relatives, rather than the narrow white bed on which

I had seen him perched cross-legged, like a tiny bird."[5] And so his grandson had brought him back to his people to die—back to the bush, where his life, and his dying, made more sense. It was there, in an Athapaskan hunting camp in northern British Columbia in 1964, that my colleague witnessed the majestic orchestration of this obscure man's death.

Like most people in such circumstances, the old Indian proceeded to put his affairs in order. Strange affairs, to Western sensibilities. In the days before the fatal attack, Japasa was heard singing to the wind and rain. Both were his friends, friends from his youth, when they had appeared to him as people in the course of his vision quest. To a boy of nine, out alone and intentionally lost in the bush, seeking an encounter with something of nature, the wind beings and rain beings had taken pity; they had come and put their hands in his, promising him their powers for the rest of his life. Power through song: they taught him how to reach them in song. Now that the old man had no further need of wind and rain power, he sang them to his bedside one last time, to tell them he was soon to die and they could leave him now.

Likewise he summoned the foxes, his medicine animal, to give back their power as well. For foxes, he said, had appeared to him, too, in that boyhood dream-vision. One night by the fire, Japasa spoke softly of the encounter—so vividly one would have thought it had happened yesterday. He recalled how a pair of silver foxes had found him and taken him back to their den, where he had lived with them and their three pups for several weeks. As he spoke, his son leaned over and whispered an English translation for my friend's benefit: "They brought food for my dad, too. They looked after him as if they were all the same. Those foxes wore clothes like people. My dad said he could understand their language. He said they taught him a song."[6]

In the course of the story, Japasa sang that very song: the boy's song, his power-animal song. Fox song. In singing it now, for the first time ever in the hearing of other people, the old man was releasing its vital energy back to its original owners, the fox people, his friends. A giving back of power held in confidence for many decades. The next morning there were fox tracks around his tent, while, to everyone's astonishment, a moose was spotted a short way off in a clearing, easily within gunshot range. "People said the animals were coming around the old man to say goodbye to him."[7]

My wife and I have a friend, a physician, who worked for a while among the Cree in northern Quebec as a general practitioner. He remembers how one winter the snow was late in coming, and the effect this produced. The tiny Indian community became strangely agitated, yet he couldn't quite figure out the connection. Finally, he asked. His informant seemed to have difficulty framing the answer; the matter seemed so transparent, it was hard to imagine anyone's missing it. As it turned out, they were worried about the small, burrowing creatures, some of whom would die without the snow's insulating effect.

Both of these stories are relics, as it were, of a conception and discourse on nature whose full antiquity is impossible to gauge, though archaeological evidence and the universalness of these sentiments among hunter societies would argue for a paleolithic origin. What is particularly remarkable is that beliefs and practices of this sort have maintained their appeal and vigor into modern times, despite having formidable cultural forces marshaled against them. To Western sensibilities, long since purged of such wildlife intimacies by the demystifying pressures of the Agricultural Revolution, Judeo-Christian teaching, rationalism, and the commercialization of the earth, such stories seem charming, and remote. They strike one as improbable, on the one hand, and impracticable on the other. Even a little frightening, if one puts one's mind to it. Talking or singing animals, not to mention singing wind and rain, are for us the stuff of fantasy.

I am reminded of the all-so-human, genial creatures of the rolling English countryside conjured up by Beatrix Potter, Kenneth Grahame, and A. A. Milne, a world readily, fantastically open to communication with ourselves. As I write this, I wonder when I lost faith in that sort of world—probably a few years before Japasa was having his final encounter with the foxes and the elements. But, then, I doubt that this Indian would have put much faith in Peter Rabbit, Renard the Fox, Ratty and Mole, or Winnie-the-Pooh, all of them bucolic, sentimental fantasies created in comfortable parlors. Each of them mirroring human ways, character, and foibles; all of them compelling as people, to be sure, though inauthentic as creatures. Japasa was looking for something genuine, which is why he went to the source, alone and vulnerable, without protection and without sustenance, renouncing hubris. The ultimate supplicant. In the spirit of an ageless American Indian tradition, Japasa sought empowerment from something explicitly other-than-human, in the firm conviction that one's humanity remains incomplete and unhinged without it.

Six years after Japasa exercised his power songs for the last time, singing them back to their wild origins, I began what has turned out to be a spiritual journey of my own, hoping to understand the terms and potency of that relationship he so perfectly embodied. I didn't know his story then; it would be many years before I was to hear it. In 1970 I began an inquiry into the nature of the relationship between the Indian and the land in North America. I had just transferred into graduate history from a program in graduate biology at the University of California, merely by walking across campus and asking the history department if it would admit this disenchanted molecular biologist. It did, with justified misgivings.

I suppose it was my biological bent that sent me immediately into the esoteric realm of American Indian natural philosophy—my scientific background plus the narcotic of the environmental movement, then in full bloom. American Indians were the height of fashion in those days among environmentalists and hippies; they were applauded for their nature wisdom and inherently conservationist

ways. Then, as now, there was much nonsense, more wishful thinking than fact, being voiced on the subject by whites and even Indians.
At the time, I found the slogans and bromides irritating, and published my irritation, arguing from historical evidence that the issue
of the Indian-land relationship, as we might call it, was far more
complicated than a blithe enshrinement of the generic, ambered Indian as some sort of ecological guru. My initial article created an
uproar, even drawing an approving, albeit misleading, editorial in
the *Wall Street Journal* (to my embarrassment).[8] Various popular and
scholarly broadsides followed.

In the years since, I have continued to brood over the issue, from
the perspective of historian, ethnologist, and biologist, though of late
mostly as a human being shorn of such academic pretense, sensing
himself orphaned from the earthy ways of this planet and wondering
how this came to pass. I have begun to suspect that the animal connection—or disconnection, as it now more truthfully is—may hold
a key to understanding my own vague, perhaps distinctively male,
sense of dissociation from the natural world. That is why Japasa's
story and the numerous ones like it I have encountered in ethnogra-
phies of American Indians draw me so powerfully, though I confess I
do not understand their full import, and doubt anyone does. Still, I
have witnessed enough stories to have become convinced that Japasa's connection with the world through animals is a form of ex-
pressing a universal human urge. More than an urge: a necessity.
Beatrix Potter, Kenneth Grahame, A. A. Milne, C. S. Lewis, J. R. R.
Tolkien, Joel Chandler Harris, and the countless other spinners of
animal-human tales that have entertained, instructed, and even
frightened children over the ages, all spoke to a biological connection which has never been bred out of us. The toy animals nestled in
the crib, the animal pictures adorning the child's first book, the pets
our children crave (as do many adults): animals are still "good to
think," indeed, essential to think.[9] Perhaps more than anything else
in the world, they guide us in understanding and accepting ourselves

and other people. Animals are fundamentally comforting. "The mind is made out of the animals / it has attended. / In all the unspoken languages, / it is their names."[10] For Japasa to retrieve those names from his subconscious and give them consciousness, he had to learn their songs. Fox and wind and rain song, in his case. Each communicated its essential powers in song—a concept I find intriguing.

∿∿∿∿∿∿∿∿

Without a doubt the most remarkable feature of hunter language and artifice in our own time—and it was presumably true for paleolithic times as well—is its animalness. Over the past five hundred years, ethnographic inquiry has repeatedly turned up this phenomenon. We see it in the Ice Age fauna adorning cave walls at Altamira (northwestern Spain) and Lascaux (France), indeed in art-speech (we might call it) pecked and painted on rock surfaces the world over. At many a museum you will find animal-visaged fishhooks, spear and harpoon tips, knife and spear and club handles, or the straightforwardly animal icons collected from hunting societies globally—crafted of bone, wood, ivory, stone, tooth, antler, and metal. It is the awesome intellect of that hairless biped, *Homo*, imaging, in artifice, the form of the other-than-human being out there and flexed within.

*Artifice* does not do justice to what is actually going on here, for these animalized creations are not inert, not inanimate, nor mere representation. On the contrary, they are believed by their creators to embody the power, the being, the life of their fleshy (corporeal) counterpart. Hunters maintain that all of nature is empowered—is conscious, intelligent, sensate, and articulate. They have drilled this into anthropologists for years. To render a creature's likeness in another substance, be it bone, wood, stone, clay, ivory, or what have you, means to inject its personhood and powers into that substance as well as to confer upon it the power inherent in that new skin. Hunter-gatherers conceive of themselves, of *Homo sapiens*, as the consummate imagers of creation. By imaging a creature in artifact, humans consort with both the creature and the substance (what we would call the medium). Such creations pulsate with animal as well as material (bone, ivory, wood, stone, etc.) power. It is *Homo*, the compulsive cosmic artificer, who has the unique anatomy and mind to express these other life forms in interpenetrating, articulate, conscious imaginations (images). Hunting peoples seem convinced that this is what humans *do* (and are to do) with those marvelous hands of theirs, stimulated by that ceaselessly (even in sleep) imaging mind. Putting it in my terms: this is humanity's anatomical and neurobiological trick.

Herein lie the real power and beauty, and deepest meaning, of all human artifice: in its powers of association, of connection. Artifice enables the craftsman to flow into another shape, another place of being; operating as a key, it unlocks the door to the realm of another being, with which the person merges. Artifice thus assists in setting the terms of selfness. Initially, I believe, this was the special property of artifice: making more vivid, even more numinous and wondrously terrifying, the realm of true humanness, of true self. For, in primal artifice, if ethnography is a guide, all was magical, artistic, and creative, connecting its wielder and creator to the world about. It worked, however, only as long as mankind could control its immense capacity for hubris and fear, and remain cognizant of the metaphysical hazards of adopting a foreign technology (as happened when Old World met New). Let us consider taking the genesis of human artifice out of the realm of the imperative and crudely functional into the realm of the magical, the aesthetic—but above all the communicative.

Throughout our evolutionary transit, obtaining food and shelter and, eventually, clothing were mundane tasks that had always worked, so long as society acknowledged death as acceptable and kept the pressures on the landscape modest. "Necessity is the mother of invention" may well hold for neolithic peoples and their spiritual heirs, but it does not necessarily work, I believe, for paleolithic so-

cieties and their antecedents. It is revealing that the lure of European
trade goods to aboriginals around the world was at first aesthetic and
spiritual: an aesthetic and spiritual communication, with the place
itself and with the bearers of these novel tidings. The "savages" knew
they did not require all of that European hardware to manage their
lives; the initial appeal was hardly practical, in the usual sense of the
term. Historians record that many an object of strict utility was
cannibalized or otherwise redefined in the early years of contact and
sported anew as adornment. Copper pots were cut up into earrings,
axe heads worn about the neck as pendants, and so forth. Such articles bespoke a newly expanded context of magic, of power, of
humanness—a subtly and, as it turned out, dangerously changing
definition of humanness in relation to place. The wholesale incor-
poration of such alien artifice was tantamount to speaking an alien
tongue, with attendant realigning of spiritual power relationships.

Homo's other great trick is speech. For hunting peoples, the
fundamental principle has always been to learn the language (which
is Power) of the other-than-human being. One way hunters do
so, though not the only way, is by projecting themselves into the
spiritual-physical space of other-than-human persons (animal or
plant) in the ecstatic experience known to ethnologists as the vision
quest. As in Japasa's experience, boys are coached for years in the
discipline of solitary fasting and praying, to prepare them for the vi-
sionary encounter.

> One summer in a wood, he was to examine his ambitions and prostrate
> himself through ritual fasting, thirsting, sleeplessness, fierce con-
> centration on the goal of "seeing and hearing" a well-disposed manito,
> weeping and blackening his face with charcoal from the fireplace and
> rubbing ashes on his hair. He recalled all he had heard about "power,"
> in his ears were the pleas of grandparents and parents to "make some-
> thing" of himself, "to fill [his] emptiness." To the manitos he declared
> his pitiableness, his utter dependence on spiritual aid, the desperate-

ness of life. Through the miasma of hysterical fear there came to the fortunate seeker—dry-mouthed, empty-bellied, and light-headed—a kindly spirit voice that assured, "My grandson, here is something with which to amuse yourself."[11]

Girls followed a parallel regimen, although their power experience, when it came, tended to be more of the earth, while the boys' was of animals.

Visions were vivid engagements with potent earth beings in suspended time. The Visitor taught the child its song. Nature, for hunter-gatherers throughout recorded time (we imagine this held true for paleolithic hunters), is, above all, song. Power songs. Songs with flattering words; they are always songs of courtesy, conveying care and respect. "Because he who comes looks so fine" was the caribou lure song of Jérôme Antoine, a Montagnais-Naskapi from the Quebec-Labrador peninsula. The old man had dreamed it a long time ago, Frank Speck was told in 1925 by the hunter's grandson. Antoine had sung it when courting (hunting) caribou persons over the years and then, evidently, had taught it to others in the community, who continued using it to commune with, flatter, and court caribou after his death.[12] As it happens, the song has an old lineage among the Indians of this region of Canada. James McKenzie, agent for the Northwest Company, recorded a nearly identical version well over a hundred years earlier in the chant (sung, he said, "in a sort of ecstasy"), "He comes—He comes—I see him, I see him—he is dressed very fine."[13] The details of its survival are not of much interest to most of us now. What is important is that man-the-hunter realized that animals spoke languages human persons were capable of learning, on animals' terms and in their realm—so man-the-hunter imagined. At least he could learn some essential ingredient of the animals' songs, as Speck and McKenzie vividly witnessed.

Women in hunting societies tended to have a complementary revelation at first menstruation, being then made privy to the earth-plant power of creation-procreation-sustenance which they housed within themselves. They, in turn, would learn the lyrics and music of plant beings, who were by definition medicine beings.

As with artifact, so with these learned songs: each is alive with the power of its other-than-human source and represents a courteous communion with that being. Tempering this is the knowledge that, in recent times, anyway, hunters have occasionally, if not habitually, taken an inherently powerful object and purposely desecrated it, to offend the spirit-keepers and, through their retaliation, produce a desired end. This anomalous strategy is most commonly used in weather control—weather magic, about which I will have more to say later on—where the usual aspect of courtesy is inverted in the service of what appear to be strictly human purposes.

One further point worth considering is that even everyday, mundane speech is cautious, since speech is by definition powerful. It is creative, reaching beyond ourselves. Throughout recorded history hunters have been very circumspect in what they say, believing they are being heard by a sentient, conscious universe—a gallery of intelligent beings who, if offended by injudicious words (ridicule, bragging, undue familiarity, profanity, etc.) can take reprisal, usually by a steadfast refusal to be taken as food or by inflicting disease or doing other violence. Consider walruses. "Walruses are like people," the Eskimos say. "They hear you when you talk and if you brag they might get you for it. If you never hunted them before," cautioned one hunter, "I'll tell you something now. Remember when you hunt walrus you must not act like a man. Do not be arrogant; be humble. Always respect the walruses and watch them closely when you hunt them."[14]

Since animal and plant beings eavesdrop on our conversations, it behooves us to refer to them respectfully, if we wish to cultivate their favor (i.e., if humans still wish them to yield themselves). Even myths, as sacred narratives from the beginning of creation that re-

*[margin note:] perception / reciprocal relationship between human and nature*

mind each new generation of the ordained relations between all sentient beings, as stories that explain the course and nature of life from the archetypal, original episodes—even these are thought to be alive and highly charged. Many have a seasonality to their telling; speaking them out of season is to summon their still-living principals (originals) out of context—which offends them.

It's not so much that something is conjured up in speech and art or artifice as it is that connections, linkages, or hinges are made between things, or better yet, between beings. There is a handshake. Mythology maintains that all of creation existed first as thought, which was then uttered as speech, by which powers the primal beings took shape (or were given space in which to function). These primal beings, including animal beings, among them the notorious trickster-transformer, created in turn by manual artifice—again, so the stories say. Hunters think of humankind as the keepers of these formulaic stories, the narrators and symbolizers of the blueprint of creation. They believe themselves responsible for repeating these tales in order to keep them alive and, further, to regenerate the system. Mankind, in fine, has the mind uniquely capable of imaging the vast yet interconnected network of creation and rendering it in language and material structure. In this sense, hunters view themselves as the historian-regenerators and artist-regenerators. This, for them, is the great calling of our kind.

〰〰〰〰〰〰

All of this is strikingly different from the popularly held opinion of so-called primitive societies. The common assumption is that hunting and gathering is by definition a desperate, brutish, hand-to-mouth way of life. We envision a mind obsessed with satisfying the gnawing urges of the belly—and when not the belly, the libido—in an often difficult, unforgiving, God-forsaken landscape. Or we visualize an Edenic landscape where savages squandered their days in idle sport, sexual license, and intellectual torpor. One can pile up image after image, none of it flattering. Western travel literature, from the calculating Columbus to the vitriolic Mark Twain, is loaded with it.

Roughly twenty years ago, cultural anthropologists began systematically discrediting this then-prevailing dogma about hunting societies, as careful research began to reveal that these "primitives," insofar as they were able to live largely free of the accouterments, vices, and promptings of civilized society, were not (and had not been) economically destitute after all. Destitution resulted from contact with the West; it was not aboriginal. The picture that emerged during this revision was of a people who, when left to their own devices, were typically well nourished, were noticeably free of anxiety about the food quest, and who, far from spending every waking moment occupied with the necessities of life, were blessed with an abundance of what we would call leisure time. These revelations moved one prominent anthropologist, Marshall Sahlins, in a now classic phrase, to pronounce hunters "the original affluent society."[15] And the tag has pretty much stuck, under repeated scrutiny. It began to appear that the romantic myth of the "Noble Savage" (most memorably expressed by a series of French essayists) might not be so mythic and romantic after all.

The upshot is that anthropologists, save for a few die-hards scattered about, no longer regard hunting and gathering as having been on the whole a risky business. It is journalists, Madison Avenue advertisers, and cartoonists who seem impervious to the message. At least hunting is not thought to have been precarious under normal circumstances. Only when the system was stressed in certain key ways did foraging become problematic. The stress might come from either the human side or the earth's side of the equation. In the latter case, the crisis was more than likely perceived by man and woman as something beyond their control, perhaps even a punishment of some sort. One gathers this by examining the dynamics of hunting cultures both in the ethnographic literature of this century and in the historical record of the one hundred to four hundred years prior to that.

My aim, however, is not so much to review the recent academic wisdom on hunter-gatherer economies as it is to probe the meaning and purpose of discourse and artifice in such societies. It is my contention that speech and artifice were the most interesting elements in mankind's repertoire when modern man, *Homo sapiens sapiens*, appeared roughly 50,000 to 100,000 years ago (closer to the former date in Europe and Asia, and to the latter in Africa and the Levant). A brain of formidable imaging and syntactical capacity now fired a vocal tool (tongue, lips, and larynx) and grasping device (fingers and thumb) of commensurate skill. Presumably this wedding of intellectual (linguistic, perhaps above all), sensory-motor, and anatomical skills amounted to an improvement (risky word, that) over the neurological and anatomical equipment—and hence capability—of the more archaic hominids (*Homo erectus*) which fully sapient man began replacing in ecological niche after ecological niche in Europe, Asia, and Africa. Although this is utter speculation: What we see are more numerous and, to our eyes, more sophisticated and interesting artifact assemblages and rock-wall paintings by the rapidly proliferating new sapient type.

Bear in mind through all this that mankind had always, throughout our prolonged and tortuous career in a variety of creaturely guises, been a forager—a hunter of other game, a fisher, shellfisher, and gatherer of wild plants. A hunting and gathering economy—*spirituality* is a better way to put it—is the oldest game known to humanity, known and practiced by our evolutionary ancestors as far back as we dare go. Speech and manual dexterity under fine and elaborate sensory and motor control and linked to possibly new realms of conceptualization—these, it is conjectured, were the new instruments in that sacred game. Yet it must be confessed that physical anthropologists know precious little about Neanderthal and even more remote hominids in these matters.

Be that as it may, the universal image of such oral and manual discourse was one of intimate identification, indeed kinship, in a mythically literal sense, with the rest of earth's society—a speech and technology of power, though not so much a power *over* other things as the power *of* these other sentient beings. Humans learned their *human* powers and abilities from these other-than-human beings. They strove zealously to do so, in order that humans might fully converse with these other beings, and, in so doing, keep the world running properly. Hunting, fishing, shellfishing, plant gathering, tool making, house building, the making of clothing and adornment, feasting—all were part of that conversation. Not merely a human soliloquy, human speech was a more truly primal discourse, first established in the great vision of youth-entering-adulthood and then indulged in and further plumbed for the remainder of one's mortal career. "Because he who comes looks so fine / He comes—He comes—I see him, I see him—he is dressed very fine." The animal likewise heard this being uttered; indeed, the animal had taught its strains and lyrics to the attentive pupil. By learning the songs of other-than-human beings, one became joined to them—more properly, one recollected one's ancient kinship (communion) with these beings. That which we call species distinction, species separation, would have been rejected as absurd and undesirable by humans in that realm of mind and speech and artifice. Only a fool would imagine himself as somehow exclusively a *human* being. Through language and artifice one could recall and vivify the primal linkage (we might call it an evolutionary connection) to other forms of life, animal and plant. Language and art underscored *Homo*'s bestial and vegetal identity, a hyphenated identity: "I can be a frog or a fox and still be a person" (Robin Ridington).[16]

"Once we were shape-shifters," hunters remind their children, in the language of myth; today we wear the skin of human people, but through language and art we continue to commune with the creatures into which our ancestors could be transformed. Under the present dispensation such metamorphosis is thought to happen but once, in the adolescent vision, but the power of that transformation can always be rekindled in speech and artifice. Such is the power of humanity's imaging mind—when awake, when in ecstasy or trance, and when asleep in dream.

Such palpable kinship with animal and plant beings bred both respect and confidence, the other outstanding feature of hunter-gatherer conduct. "One of the most overwhelming and lasting impressions that one receives of the Eskimo hunter is that he is self-assured and competent above all."[17] For the hunter-gatherer, nature is no adversary. Quite the opposite, nature's other life forms are (to our Western eyes) astonishingly willing to furnish themselves to man and woman to satisfy obvious human needs of survival. What scholars call the *food quest* was largely taken for granted by hunters, under normal circumstances. Perhaps a better way of putting it is to say that the food quest originates in myth—a narration of authorization, of fundamental conditions—wherein an individual is reminded that he must *be* the animal before presuming to kill and eat it. A hunter can care about caribou, for example, only after becoming caribou in his thoughts, only after the transformation, as the paradigm of Caribou Man illustrates. Hunting these creatures—a dietary staple among the Montagnais-Naskapi of the Quebec-Labrador Peninsula—originated with Caribou Man, who sang the following when he came to live with the people:

> The caribou walked along well like me. Then I walked as he was walking. Then I took his path. And then I walked like the caribou, my trail looking like a caribou trail where I saw my tracks. And so indeed I will take care of the caribou. I indeed will divide the caribou. I will give them to the people. I will know how many to give the people. It will be

known to me. He who obeys the requirements is given caribou, and he who disobeys is not given caribou. If he wastes much caribou he cannot be given them, because he wastes too much of his food—the good things. And now, as much as I have spoken, you will know forever how it is. For so now it is as I have said. I, indeed, am Caribou Man. So I am called.[18]

The other-than-human persons, vegetable and animal, will give themselves to me, as long as I refrain from overexploitation, as long as I treat their flesh and substance, including their remains, with respect and avoid all other forms of offense—this is the prevailing sentiment. Nature conserves me, not I it—this is the underlying ethic. Language and artifice both explicitly and inherently confirm and reconfirm these terms of existence, this cosmic orientation, as the mind continues to image a relationship of interpenetration of the human with the other-than-human person. "You and I wear the same covering and have the same mind and spiritual strength," sing the conjuror and his drum to the game.[19]

〰〰〰〰〰〰

∿∿∿∿∿∿∿∿

*Eskimo*

Eskimo
self-conceived animal
haloed
in fur of white fox
stands
at the center of night
dismissing
silence upon silence
stars drop
the temperature of ice
the ivory point
the named
the personal weapon
shivers
waiting to unite
with the bursting heart
of the mild-eyed seal
whom the man
loves enough
to kill
devour
honor
as in a sacrifice.[20]

# 2

# *"Creation of a margin"*

〰〰〰〰〰

## *Succumbing to Fear*

Wandering through a gravel wash in the high Sierras, Loren Eiseley recalled years later how he had chanced upon "a polished flint from the hand that dropped it / ten thousand years ago. / Where is the hand now? / In what language / is the flint remembered?" That finely controlled fabricator hand, inspired by an enabling image, and given voice in some long forgotten animal-man-blade speech—words and hand now "both lost on the air."[1] What remained was this lithic mind-print alone. In what language is the flint remembered? In my own, Eiseley's artifact is but a percussion-chipped and pressure-flaked Paleo-Indian blade of utilitarian interest; its identity is practical, functional, strictly a device of human cunning and ingenuity in the ageless and necessary pursuit of sustenance.

What is interesting, here, is how I wrap words around the subject, trying to locate it. I am tempted to call it an ancient artifact, but I resist. It is here, now, in my hand, bearing a true and always present message: that our relationship to place and animalness is intimate, potent, and tricky. But I don't have the language to truly convey the richness of what my still-paleolithic mind senses; mine are words forged in a smithy far different from that known to Eiseley's ghostly Paleo-Indian mechanic and his metaphysical, hunter-gatherer heir. I speak a language long since purged of authentic animal and plant and elemental resonances and harmonics; I speak the language of the neolithic, the triumphant tongue of man apart, the voice of history. I admit to being ignorant of indigenous and possessed words, here; I do not dream or sing of such things as the individual who imagined this blade surely did. I am not "a part of it," just as this blade, when put under glass and labeled in a museum of natural history somewhere, will also become unhinged from its true constellation. A museum is not a place to know this thing, just as a zoo is not a place to know animalness.

∿∿∿∿∿∿∿∿∿

It was a place given over to winds; winds seemed in control of what transpired here now. Before winds, it had been water, tumultuous rivers of silt that over the eons had sandpapered this slit into the high plateau, sculpting it into fantastic shapes and eddies. Moving always deeper into the earth, as though to lay bare its powers. Here humans had journeyed and set down over the millennia, during the age of water; the Anasazi, Ancient Ones, had built their eerie cliff dwellings farther up the canyon, downstream, as it were, from where I entered it. Upstream—although the bed was bone-dry when I visited—the canyon stopped in a box. Dead end: the sandy, pebbly riverbed was strewn with empty bottles—of Garden Deluxe, mostly, the cheap California wine popular among those contemporary Navajo who can no longer be *a part of it*. A canyon redolent with meaning: despair at one end, faith at the other. Cosmic connection and disconnection. Both ends demonstrating mankind's powers of imagination, and that the world for *Homo sapiens* is always a potent place, a trickster's realm. All situated in the river of time, except that the river, and somehow time's quicksilver, had vanished. The canyon seemed engaged in a transaction which my culture had ill-equipped me to divine.

Chiseled into a transverse ridge on the arroyo floor was a row of figures. Petroglyphs: a hope, a dream. A communication: rabbit tracks, hooves, and a horned serpent. Humanity beseeching the powers of place for rain (for the river to come again), for game. The sacred game etched into the land, at just the correct spot. I knew it for the place of power and illumination that it was. And so I walked softly, even while realizing that I was bereft of effective communication with it—communication in words and artifice, if not in imagination. My deepest impression was that I was intruding; lacking conversation in these matters, I sensed I was out of category. I felt that what language I did control and what artifice I habitually practiced were hostile to what was going on here, to its properties. I felt watched. I had noticed the carefully piled stones and other signs of recent human meditation under a cliff overhang a short way off from the glyphs. Some contemporary of mine, Navajo no doubt, was still fluent in these powers. That pleased me, both as counterpoint to the smashed bottles and as repudiation of my culture's turn of mind, of phrase, of hand.

Still, I knew I didn't belong; on the reservation, it was a place that had not yielded to the denaturing forces of my civilization's speech and artifice. I knew I was being observed. The words of the magician snaked in and out of my consciousness: "It has become clear to me that perception has to be understood and recognized as a reciprocal exchange. When we see things we are also being seen by them. When we hear things we are also being heard. Perception is a type of communication that precedes language."[2] I liked that; I felt sacred; I felt I was communicating on the farther side of language, the place where language is born. There was nothing for me to say, audibly. Nothing for me to fashion, as the Navajo seeker recently before me had done; my hands were skilled only in the service of strangely alien errands. I stood in a place of palpable powers of earth, an intruder, and was deeply ashamed and, yes, frightened. This was a place of enormous meaning to me, and I felt lost—yet, in some ineffable way, felt I should stay.

So, staying, I knelt down on top of the smooth bedrock, slowly tracing its figures with my fingers and hands, feeling for their grace. "Traditional magicians seem to be able to flood the world which surrounds them with their imagination, to let their imaginations out. . . . If we have a creative imagination then the world itself must be filled with imaginations, not just human imaginations but the imaginings of all the other creatures around and the plants and the trees."[3] The day before, I had played the piano in the mission church in the village nearby, hymns to a Middle Eastern sky god who seemed to me almost ludicrously out of place in this landscape. The minister had heard that I could play, and implored me to. Having performed while a believer, for my dad's services many years in the past, I would do so now as an unbeliever to humor this man, the missionary who could laugh that he spoke no Navajo. Fingers that could play hymns to an incongruous neolithic deity now sought to know the song of these pecked images. All I heard were the sighs and whispers of windsong here and there, fingering the canyon walls around me, ceaselessly sculpting minarets and flutings into the pliant rock. I thought of words as sculptors of space, as turn-keys: liberators or jailers of the powers about us. The wind was certainly a better musician than I, releasing powers to which my culture was oblivious. Perhaps because I am something of a musician myself, and a canoeist, I have always been attentive to winds.

I had hiked this canyon two days before with my wife, Nina, and another couple, and on the way back she had suddenly stooped down and plucked a still perfect arrow point from the sea of pebbles and smoothened stones that form the canyon floor in the vicinity of the glyphs. The four of us spent the next hour scouring the area for more, in vain. Alone now, I was in a far different frame of mind; I was in search of a theophany which, from the previous visit, I detected was still here, and might, I halfway wished, teach me something. Might speak or sing to me. It was the closest I have come to a vision quest. "In order to touch those, I believe it's necessary to let out our imagination, to breathe it out, to cough it up if you will, to take all the work we've been doing on our interior psyche, all our spiritual work, and breathe it back into the space that surrounds us, and then we'll find that that is in fact where the psyche lives. . . . That is the magician's secret, that the spirits are right there if you would only recognize them."[4] Here and there in the distance a dust-devil coalesced and danced. In the fullness of time. "After working this way for many years I have begun to get a sense that perception is a very active and engaging work by which we constantly move out of ourselves through our eyes, our ears, our mouths, into the world and dance in that world, and shape it."[5]

Perception: being seen and heard by what I saw and heard. The idea suddenly snapped into place, into a place that fit it: here in this canyon. My sensations of being watched, of my trespass, of unease, were the reciprocal exchange my magician friend had been talking about. I straightened up and stepped away from the petroglyphs, idly casting my eye over the pebbly floor, wondering whether there might be another projectile point lying hereabout. And as I did, I had the gathering realization that such a thing would find me, but only if I imagined myself into its flesh first. I found I could let my imagination out and into this object; I could become a part of it. And there it lay, at my feet: white and flawless save for a small chip at the tip. I picked it up and welcomed it; we seemed to be two sentient beings joined.

Just as I cradled it in the palm of my hand I half noticed the faint wind of a dust-devil gathering behind me. In an instant there was a terrific report, like the sound of a gunshot immediately at my back, and I whirled around. There before me stood a whirlwind being, in stature, as best I recollect, slightly taller than me. Confronting me. The Visitor? Who can know these things. I felt I was being interrogated; I felt guilty and vulnerable holding this lithic in my hand. Was

this the Keeper of the Artifice? The Spirit of Place? I had deliberately
squeezed my imagination through an aperture and entered the spiritual dimension of this canyon; I had willfully asked to be found by a
sacred object—and had been. Was I to be surprised at finding myself
face to face with a wind spirit? I wasn't, yet I was. I had been un-
prepared—that was it. What I said—yes, I spoke as best I could,
a word and gesture of respect—seemed to come forth of its own volition. And with that the apparition whirled away to the rim and
vanished.

I tried to stay, to dismiss it and carry on with what I had been
doing. But what I had been doing was done. It was over, metaphysically. There was an immense, even crushing, sense of closure. I
knew I must go; I had the sensation of being ejected. Buttoning the
arrow point into my shirt pocket, I quickly walked out. Had the canyon closed up behind me, I would not have been surprised.

My mind raced as I drove back to the hospital compound where
we were living. I knew from ethnography that visions were, properly,
deeply troubling experiences, that they were a beginning of knowledge, of new identity, and of empowerment. I hoped this experience
would be so for me. There was also the queer sensation that I had had
no choice in the matter.

In the hours and days that followed, my rationality pecked away
at my recollection of the experience. One can guess the gist of the
debate, the insinuating logic. But I silenced it. The white projectile
point meanwhile lay on my bureau, and I would often feel it. It
seemed to be feeling me, too, with its imagination, and I did not forget what the magician had said. I now sought advice. The missionary
had lived here for some years and I knew from talking with him that
he was conscious of the spiritual sensibilities of his tiny flock; it
occurred to me that he might have heard talk of wind beings—it
mattered little what the occasion might have been. I sought him out.
Ah, he assured, he was not surprised by the incident; there were demons inherent in this landscape; he sometimes sensed them in the

very sanctuary even as he preached—it was as though my father were speaking. The explanation, the terms, were too simplistic and were incongruous to this place; they smothered something of immense potency and meaning. They were the words of a man who could not bear to give himself, sensually, to the place. The preacher was ignorant of the matter.

The head nurse at the clinic had lived on the reservation, in this place, for years. An artist, a healer, a musician and poet, she was, she told me, descended from gypsies. And I believed her: she was witchy, remote, and wildly sensual. A dawn runner, a woman who could dance stripped to the waist at night before a fire, throwing shadows on canyon walls. I described the encounter in the canyon. She replied, after a moment's reflection: "Out here, one sees things with greater clarity." Clairvoyance. That was it. I now had a word to give it, feel it, locate it. I went back and held the point, a thing not harnessing wind but born of it, and contemplated clarity. Once touched by an encounter, meaning can take shape. And it was.

We were to leave soon; Nina, in the final year of medical school, was being re-posted to the Indian Health Service Hospital at Shiprock. I am not a collector, in fact, quite the opposite. Even so, in this instance I had the impression that one could not even hope to possess this thing of that wind canyon. It belonged out there; it made sense only there, on that holy ground where I had stood; it was part of place. So, shortly before leaving, I took it back to that wind place; spoke to it, thanked it for its lesson, caressed it, and returned it. One feels foolish revealing such rites. But I knew it was right. Intuition, maybe.

〰〰〰〰〰〰〰

Human speech and artifice obviously no longer encompass the kind of communication described above and in the chapter preceding. The society (I prefer the term *geography*) of human discourse has narrowed from human/other-than-human to human/human. The other-than-human persons, both animal and plant, have been dis-franchised—defined or spoken out of discourse into dumb brutes or unconscious vegetable matter, each depersonalized by man the cosmic orator, the name-caller. In parallel and related fashion, our artifice has likewise become deanimated. Spirits are no longer seriously, truly thought to dwell in material items of human craft; they no longer animate the objects, make them function, make them go. Yet we are fond of naming manufactured things, especially vehicles it seems, after creatures or even natural phenomena whose qualities we half-seriously imagine we have ingeniously approximated. These associations, though cute or inspiring, even exhilarating, are full of mischief within the larger sphere of consciousness and performance (and, one might say, of truth) that *Homo* has always operated within and that hunter-gatherers have tended to keep firmly before them. It was neolithic peoples who cooked up a different, and unsustainable, economy of place and place-beings, in part through this very process itself: of associating, perhaps by name, an artifact of human ends with something of earth's ends—a clever, though eventually fatal, confusion. We should have the integrity and courage to recognize in this the origin and, today, perpetuation of a falsely ordered world.

No, animal names should be scrupulously left with them, not attached to battleships, automobiles, or missiles. Animal traits, on the other hand, should be invoked with extreme care. Out of courtesy, so a traditional Cree, say, or Aborigine might caution. I would echo that, while adding that to do so is to respect the accuracy and truth of what is transpiring on this earth in the largest, most complex sense. Cree and other hunter societies know that human disregard for accuracy and truth surely invited misfortune, even madness—prophetic words which may have come true, in an eerie sense, in our neolithically arranged era.

I speak in particular of Western civilization, which began this gradual and subtle process of exorcism, of demystification, thousands of years ago. I like to think of it as a swelling chorus of denunciation, of defamation, perhaps even damnation of these other beings over time. Chronological time was a powerful conceptual device in bringing this disfranchisement to pass.

We find ample and vivid illustration of the process in the Judeo-Christian Old Testament: I am the only true God, announces Jehovah; all others are fraudulent and ineffectual. Moreover, this Judeo-Christian deity is no provincial, but is sovereign throughout both the heavens and, more urgently, the earth. At the same time, mankind is his special creation, existing several orders of magnitude above the other creatures and, even more, above plants, occupying a vastly superior category of being. For man and woman alone possess a soul.

The great virtue of Judeo-Christian man and woman is that they can converse with the deity and hence learn the god's will and terms of life. It was this which would ultimately distinguish mankind from all other kinds, the ipso facto lower kinds. The message from the deity was one of eschatology, of an accounting that went beyond one's present mortal phase into the eternal world of spirit beyond: a language of divine judgment and, linked to that, salvation. Thus, the individual chosen by the Christian god was to speak to (not be spoken to by) the unsaved races throughout the world, to communicate the latter's inherent depravity and, that registered, the way to salvation through faith in Jesus Christ. For the good pilgrim, the only other person to be engaged was the unregenerate, spiritually blinded, yet potentially salvageable heathen. Animals and plants were beyond the pale—scripture, the early Church fathers, and even Aristotle, were clear on that. Such creatures were tools, were larder, were real and enumerated wealth and property (chattels). There was to be no godly revelation for their minds and no elevation. Not being fashioned in the deity's image, they were innocent of both intellect and that divine substance, soul. Although through the centuries there have been those who have argued that animals are endowed with intellect and soul, theirs has turned out to be a minority and sometimes even persecuted view. In place of intelligence and divine spark, animals were said to possess something we moderns nebulously call instinct. Mankind had complete license, courtesy of the god, to conquer these creeping, quadrupedic, or brachiating things; the hissing, snarling, chattering, moronic forms or totally inert vegetable matter of a cursed nature, which had been created for our use and enjoyment.

Shifting to the other major inspiration for Western culture, the Greco-Roman, we find an early pantheon of Olympian deities gradually being transformed into increasingly less credible and less relevant personages which ultimately wind up as objects of jest and derision. For its part, the Roman state would take its cue from the divine will (and eventually, when this too was unmasked, from the armies) of its emperors. Outside the polis and the civitas was the barbarian Other; inside, sheltered by the wall (Romans being notorious for attempting to separate the two by stone walls), was the civilized Self. Animals and plants might be spiritually animated, especially in olden times, but it was more typically and increasingly a case of field, woods, and stream being home to spirit creatures— essentially monstrosities and marvels—of human fantasy. Not real, as in the hunter-gatherer sense. In time, Christianity and Moslemism would sweep the vestiges of Greco-Roman pantheism into the dustbin of history—the juvenile realm of imagination and fantasy—and ride triumphant (more so among the elite than the peasantry) in the lands where Zeus and Jupiter had once reigned. The point being that animism, of the sort described above for hunters, would languish and, with Biblical and Koranic imperial hegemony, eventually be sorely persecuted (though never totally eradicated) in the parts of the Mediterranean world where Greek and Roman cosmologies had once held dominion.

I am acknowledging what scholars have recognized for centuries, to the point of its being cliché: that the intellectual and ideological contours of Western civilization were derived mainly from these several Mediterranean traditions, the Judeo-Christian-Moslem and the Greco-Roman (the Greek more since the Renaissance than before). It is true, too, that each of these traditions has taken pains to disarticulate, in all senses of the word, the animal and plant realms, along with a substantial portion of the human race, which was rendered as "barbarian" and "savage." Again, I have given but a fleeting and fragmentary glimpse of that process of thought and action, while at the same time I have said not a word about the other great agrarian religions, especially the Eastern ones, and the subtle or overt ways they, too, have asserted human primacy over animals and plants. "Han Yu's Address to the Crocodiles," at the end of chapter 3, illustrates the process in Confucian practice. My focus on the West is prompted, I confess, by the accelerated enveloping of the global human community in Western ways since the sixteenth century Age of Discovery and Exploration. Today, the tempo of that process is mind-boggling.

The fact is that the process of disarticulating nature reaches back much further than any of these cultural traditions would have it, all the way back to the environmental stresses and oral-artifactual responses of paleolithic hunters. The Greeks and Romans, with their demystified philosophy and history, their emphasis on the chosen individual, and their civilized-barbarian dichotomy, and the Hebrews, Christians, and Moslems, with their extraordinary violence, chauvinistic god, and fanatical belief in a chosen people, were the heirs rather than the fount of a process more universal, more anthropological, and more ancient than is commonly acknowledged.

~~~~~~~~~~~~~~~

Archaeologists tell us that sapient man was a most efficient and cunning hunter, making his debut at what may have been the most opportune hunting season in earth's history: the Pleistocene. *Homo sapiens sapiens* swiftly established its kind as the predator nonpareil in that era of mammalian giants. Equipped with that terrifying weapon, fire, augmented by the spear-thrower (atlatl) and a finely crafted stone spear point lashed to an end-split wooden shaft (or fore-shaft), razor-edged knives and chopper blades (both ideal for butchering), and being withal a past master at communal tactics, late Pleistocene *Homo* advanced on the vast herds of giant mammals which roamed the extensive grasslands and savannas that characterized the age. Man would harvest these prey in a drive maneuver, deploying fire and smoke to frighten them into an enclosure (corral) of natural features, where they could be slaughtered at leisure. Likewise, fire and smoke could direct a stampeding herd over a bluff or cliff to their death or immobilization below, with the same consequences. The hurled spear, given superhuman thrust with the added leverage of the atlatl, would easily puncture the hide of a giant beast, which, in its escape or enraged charge, would simply bleed to death, or to incapacitation. So man has hunted big game down through the millennia into our own time. Archaeologists have turned up evidence of each of these weapons and strategies the world over.

There is one school of thought which argues that late Pleistocene man was in reality a superpredator, largely responsible for the worldwide extinction of numerous megafaunal species in late Pleistocene and early post-Pleistocene times. Paul S. Martin, author of this overkill thesis, observes that wherever sapient man and woman migrated over the face of the earth, penetrating continents and subcontinents and major islands that had never before witnessed this new version of *Homo*, there soon followed a paroxysm of mega-mammalian destruction and extinction of unprecedented proportions. Big game species certainly had vanished into oblivion during previous geologic ages—that was not novel. What was new was extinction without replacement by similarly adapted species. The literature on the subject is quantitatively weighty, and sometimes technically ponderous as well. The overkill thesis is hotly contested by those who feel the evidence is too skimpy or ambiguous to sustain Martin's startling proposal. These other scholars counter that the culprit was more likely climatic stress, creating a breeding crisis and so forth, while they allow that human predation was but one of a constellation of challenges to mega-mammalian survival.

It hardly matters, for our purposes, whether overkill was the chief cause of all those extinctions; it is enough to appreciate just how dramatically effective man the hunter was in the game-rich and pitifully vulnerable world of the Pleistocene. It is easy to see the vulnerability of the prey in the hunting practices outlined above. Animal naiveté must have been part of that vulnerability. Fearlessness of mankind was an evolutionarily fatal trust for many species, as it turned out. Archaic *Homo*, the fruit and nut eater cum small-scale hunter-fisher, had, in the cradle of Eurasia and Africa, become transformed into modern human, still frugivorous but now a tactical hunter of unprecedented wit and deadly force and possessed of an inordinate appetite for flesh, fat, and marrow. We witness the sad results of such a confrontation in the modern analogue of whale hunting, for that giant mammal is a trusting and whimsical soul, whose sweet songs mankind, save for a few traditional Inuit, Aleut, and Amerindians, has forgotten how to sing. There is even the temptation to wonder whether sapient man, as a result of his deadly efficiency, is really the author of animal fear of our kind (death, wrote Eiseley, "is in the clothes of men"[6]—a literal truth when humans took to wearing animal skins).

As the huge ice sheets receded for the final time, warm, moist conditions returned to the northern latitudes (the antarctic cap is of less concern here), and vegetation responded predictably. Archaeology shows the big game hunter now tempering that strategy, globally, with a taste for a broader-based diet, emphasizing smaller mammals, along with shellfish, fish, gathered edible plants, and perhaps even casual cultivation. We must remember that the rudiments of plant cultivation are largely self-evident and simple: discarded seeds in a refuse pile germinate, take root, and grow up into another useful, edible plant. Or perhaps one weeds and waters around a favored plant. Behold the dawn of cultivation—hardly the dramatic affair we commonly imagine.

The excavated record from this preneolithic (i.e., mesolithic) phase shows us a grocery list notable for its diversity. The adrenalin-pumping days of the spectacular big game were not completely over —and for some surviving hunter societies have not ended even now—though they were seemingly modulated by greater reliance on other foods, especially marine and aquatic protein and collected carbohydrate (plants). Big mammals, including bison, moose, caribou, white-tailed deer, elk, and bear (in North America, to take one example), would continue to be a major source of calories, essential amino acids, minerals and vitamins (found in the marrow and gut contents), and the minute amounts of those few fatty acids humans are incapable of synthesizing. But, again, the trademark of this emerging mentality and performance (I hesitate to call it *economy*) is diversity.

What we are witnessing is the so-called economic expression of a major shift in the world view of hunting and gathering peoples, a shift that in all likelihood was accompanied and even conceivably initiated by a change in the language and artisanry of man the hunter. On the matter of human discourse with and about the earth under this new dispensation, all we have to go on is the language of modern hunter-gatherers. And here one cannot help but wonder if the conservative tone of modern hunter speech marks a lesson learned from the slaughter and extinction of the Pleistocene megafauna: Did that fiasco (if that's what we can call it) instill a sense of restraint in mankind's approach to nature? Perhaps. But matters so remote and intangible must always remain in the limbo of speculation.

Personally, I like to imagine that the speech and artisanry of modern (i.e., sixteenth- to twentieth-century) hunting peoples is a close approximation of that mesolithic ancestral model, since we see both societies harvesting the earth using pretty much the same strategies, devices, and focus. The tool kit of these mesolithic societies is, for instance, strikingly like that of contemporary hunting societies, while it is more diversified and refined than that of its paleolithic predecessor. Projectile points tend to be smaller and more finely chipped and flaked, and there is ever more regional diversity in style (where style suggests enhanced effectiveness, though it may reflect merely difference in taste); the bow and arrow replace the spear-thrower and dart in the Americas (where I am getting most of my evidence) as the principal missile of the chase; net sinkers and fish-hooks and harpoon points become profuse; and mortars and pestles imply a greater reliance on hard seeds, nuts, and grains than before. Jewelry, bone combs, awls, needles, and the like also begin turning up in relative profusion in the archaeological record. Rummaging through these people's garbage, archaeologists uncover enormous heaps of mussel shells, certifying a consuming interest in these marine and aquatic sources of food—a taste sustained up into modern times. As for land game and wildfowl, species identification of the bones of those creatures killed and butchered has revealed them to be identical in kind and, with little variation, abundance with those still popular when Europeans first appeared on the scene several centuries ago. Nor do the comparisons end here.

Based on this kind of evidence it might be more appropriate to assign the ideology of the hunt, which I associated above with paleolithic hunters, to mesolithic hunters instead, especially when one factors in the possibility of massive overkill by paleo-hunters. Still, I suspect that animalness in speech and an attitude of confidence in harvesting the earth were characteristic in both paleolithic and mesolithic societies.

The important point is that, with the end of the Pleistocene, roughly eleven thousand years ago, and its attendant climatic (warmer, wetter), geological (e.g., inundation of continental shelves), vegetational, and faunal shifts and transformations, man the hunter adapted with a notably mixed and flexible food economy. The hyper-efficient mode of hunting that quite possibly contributed to massive extinctions was now tuned to a transformed environment where diversity of food types seemingly made most sense. I have already noted a corresponding shift in artisanry. If we can infer from modern hunting societies, such artifice was numinous and spiritually animated—I would suggest, still so (as with paleolithic hunters).

∧∧∧∧∧∧∧∧∧∧∧∧

Somewhere in here a process of truly earthshaking proportions occurred, a change whose motivation and circumstances remain frustratingly elusive: the shift from merely gathering wild food to actually engineering it into being. The neolithic. It happened in a half-dozen or so foci around the world—or at least continued in a sustained fashion in these spots—just a few thousand years after the end of the Pleistocene. Not simultaneously, to be sure: plant and animal domestication began around 7000 B.C. in the Middle East and closer to 5000 B.C. in Mesoamerica (with corn agriculture), to pick two of the most intensively studied episodes of this global phenomenon.

Back in the days when it was academic dogma that hunting and gathering was an inherently inept strategy for living, the neolithic was hailed as virtually inevitable: mankind's ceaseless efforts to wrest a decent living from a hostile landscape were finally rewarded with the insights of farming and pastoralism. *Homo sapiens*, perennially struggling to master the environment, had finally achieved the great breakthrough: the harnessing of plants and animals for humanity's own higher, God-ordained purposes.

That kind of anthropocentric, teleological reasoning might have been compelling when the study of history and anthropology was still in its infancy, say, a hundred years ago, and one is embarrassed to witness it well into the present century, but let it be said emphatically, that it is no longer credible today, nor has it been for decades. The question, now, is not why it took man-the-hunter so long to make this inevitable leap from a forager economy to a producer

economy but what tempted certain groups to pursue a course of subsistence so different, metaphysically, from the one they had in hand.

The question becomes even more interesting when one considers that the fundamentals of plant and even animal domestication
were doubtless familiar to mesolithic hunters. The dog, for instance,
seems to have been a source of food ("dinner dog") in various parts
of the world well before the concerted shift from a hunter-gatherer to
a producer economy. Often the question becomes, Who domesticated
whom? Did man enslave certain species or did they initially recruit
him for their own ends? Archaeologists have long contended that the
New World turkey, a pest around human refuse dumps, domesticated us, while natural salt licks would have served as a magnet for
deer, resulting in a quasidomestication of these cervids. It would not
have taken any great genius to figure out how penning and feeding
certain species would ensure their company. The problem lay not so
much in dreaming up the concept; the problem was whether one (or
the community) wished to provide provender, cover, and contain-
ment for these proximate creatures.

Right here we begin to confront the philosophical underpinnings of the neolithic. Assuming that the basics of plant cultivation,
animal care, and animal impounding were understood by mesolithic
societies, what would have moved them to concentrate heavily on
this capacity to generate food rather than simply gather and hunt it at
nature's whim? I cannot answer that definitively. No one can. Scholars have built careers on propounding theories—and burying others
—to explain the process. But in the end it is all, including my own
thoughts, conjecture. My interest lies more in ruminating on the
new ideology of nature—one might say, the new cosmology—that
was seemingly ushered in with this new approach to food-getting.

Several points merit special attention, even reiteration, before I
proceed. First, the *principles* of domestication were probably known
long before they were systematically and, it appears, preferentially
employed. Second, the actual *technology* for rudimentary plant cul-

tivation (e.g., firing a tract of land, using a digging stick) and animal domestication (e.g., impounding) were likewise in place well before the neolithic. Again, this is deduced from historical and ethnographic reports on modern hunting societies, who oftentimes have been casual and rather indifferent farmers, though it is conceivable that they borrowed cultivating tools from farming societies. What was lacking, at the threshold of the neolithic (always a gradual process, not a Big Bang), was a proximal incentive combined with a new or enabling image of the self vis-à-vis nature (the cosmos). It is this new image of self relative to plant and animal persons that fascinates me. Herein, I believe, lies the genesis of our modern domination of nature, much as the Bible enjoined: "Be fruitful and multiply, and fill the earth and subdue it; and have dominion over the fish of the sea and over the birds of the air and over every living thing that moves upon the earth" (Genesis 1:28). This wholesale and outrageous invitation is conceivably no more than an a posteriori rationalization for humanity's new posture toward its surroundings, a matter of pinning it on the deity whom mankind had invented largely for that very purpose.

〰〰〰〰〰〰

As we look carefully at the neolithic—and it matters not which part of the world one chooses for this exercise—we begin to see that several characteristic themes emerge. One of the more striking is what Lewis Mumford has called the creation of a surplus (a "margin" is how Frederick Turner puts it), be it of Old World cereal, New World maize, or what have you. "The great fact about neolithic technics is that its main innovations were not in weapons and tools but containers," asserts Mumford. A surplus of food called into being, or permitted, by the technology of the cistern, the subterranean cyst, the ceramic pot and woven basket, "bins, barns, granaries, houses, not least great collective containers, like irrigation ditches and villages"—any or all of these. Witness nature's fruits now being produced in excess, an insurance against anticipated shortfall. Perhaps produced, as well, as a commodity of exchange.[7]

A second characteristic is the appearance of the city as ceremonial center, together with an unprecedented population surge and concentration. Social stratification, along with specialization of labor, typically accompanies this novel demographic shift, and one soon witnesses the installation of a ruling elite with potent priestly powers and obligations. A margin, a population boom, and a priesthood—three developments that throw considerable light on the metaphysics of the neolithic.

A few words about priests. Priests are by definition intermediaries, intercessors, who undergo a period of rigorous training in the tenets and rituals of a cult and only then are anointed practitioners. Instruction is passed along from priest to priest—person to person. Shamans may be thought of as the rough equivalent of priests. And here we see a curious phenomenon: shamans abound in hunter-gatherer societies, while priests, aboriginally at least, are almost unheard of. Shamans are in business for themselves; basically, they are loners with remarkable spiritual transmitters and receivers. They gain their powers from intercourse with spirit familiars, not the study of dogma and ritual, although they may learn some of their shamanistic techniques or arts (e.g., the sucking and blowing and legerdemain of curing) as an understudy to an individual with greater experience. It must be emphasized, however, that among hunter societies, all individuals are expected to acquire and wield spiritual powers unique to themselves, a process that starts with the vision. There are a number of other distinctions between priests and shamans, but the really key difference is that priests are devotees of transcendent, otherworldly deities. Priests, in fine, mediate between the sky gods and humanity.

Sky gods are, categorically, creations of the neolithic. They are cosmic beings with a powerful agenda: a plan for mankind and the remainder of creation, in which a chosen people become the agent in carrying out that divine will. Notice that sky gods are conspicuously absent from hunter practice, except where hunters have since been influenced by neolithic cultures and civilizations. In fact, hunters don't worship gods at all; they don't worship anything, for that matter. They converse with local, earth- and sea-bound spirit persons without adoring them. They are animists, not theists. The communion that hunters have with animal and plant beings is at a rather mundane level of existence and discourse, and they do it themselves rather than assign that task to someone else, such as a priest. A shaman will be engaged only when one's own powers of discernment or facilitation have failed; thus, a shaman is a last resort and by no means the spiritual locus of the community. For here every individual is his or her own spiritual spokesman.

Perhaps I ought to clarify that hunters do detect other sorts of spirit beings besides those of animals and plants, vague entities associated, say, with peculiar landforms, such as waterfalls and rapids, or bizarre underwater monsters, giant man-eating creatures of the forest, and so on. One has difficulty knowing how much of this potpourri is aboriginal, predating contact with, say, the Christian West, with its assorted horrors, traumas, and outlandish tales. Nonetheless it seems certain that there have always been supercharged beings in hunter-gatherer imagination, beings who, like the potent animals and plants, are capable of inflicting serious misfortune on man and woman. One deals with such beings circumspectly, though still on a mundane and very personal level. Whatever their powers, and as dangerous as they may be, they are always of this world; they always inhabit the local geography; they are never transcendent or extruded.

The neolithic witnessed a profound transformation of this geographically local, immanent spirituality. It was as though incipient farmers took the spirits of the numinous places, joined them with the spiritual essence and personhood of fauna and flora, and hurled that whole potent assemblage into the heavens as a cosmic zygote—the reverse of the myth of human creation by the god. However it happened, from the dawn of systematic farming until now, mankind has been afflicted with gods, either one's own or someone else's.

Priests, then, are mirrors of deities; and deities, in turn, mirror time—more accurately, what we might call the errand of time. The three (priests, sky gods, time's errand) constitute a fundamental and novel element of early farming—what we should now start calling civilization, using the word in the anthropological sense of an urban society feeding itself by agricultural or pastoral means.

Priests, sky gods, a chosen people on a divinely commissioned errand through time (and often through geographical space, as well), a food surplus, and runaway population—what was it that possessed mesolithic hunter-gatherers to recast their image of themselves vis-à-vis nature into such peculiar and freighted terms? (I am referring to the first farmers, not those hunters infected with someone else's ideas.) I said earlier that scholars don't know for certain. We have a clue, however, in the production of a food surplus linked to a population explosion, although it is well-nigh impossible to determine which came first.

Before continuing, we must appreciate that one of the most distinctive features of hunter-gatherers, historically and ethnographically, has been their willingness, indeed insistence, to limit their population growth (except when seduced by the surpluses, so-called, of civilization—but that's another story). Through a combination of infanticide, birth spacing (through prolonged lactation, abstinence, and plant-induced abortions), marriage exogamy, and geronticide, hunting societies generally kept a tight rein on their numbers. Scholars now scoff at the idea that they were forced to do this because of a rudimentary technology. They were not. Hunters had the practical means to exert much greater pressure on faunal and floral sources of food and raiment, and they had the knowledge of where these sources were located in seasonal abundance—each to its time and place. The long and short of it is that they lacked the incentive; they lacked the image, above all.

Hunters, of course, move about the landscape, not randomly, hoping to encounter something delicious, but in a directed and efficient manner. Hence, they put a premium on mobility. Historic and ethnographic sources confirm that infants and the very old and infirm could hinder that mobility. Yet, as real as such encumbrance might have been, it surely formed only part of the rationale for population control. In this as in other matters, hunter pragmatics conformed to a larger constitution of identities and protocols with the other intelligent things of one's environment, and must be assessed within its terms. It is this larger *image* of living, of identity and protocol with other life forms, that I am trying to focus on and argue is exemplary from the hunter-gatherer realm.

One can continue pondering the whys and wherefores of hunter population regulation, but for the moment it is sufficient to note that this stricture was jettisoned under the farming proposition. Farming societies are expansive, in various senses of the word: farmers create a margin, a food surplus; hunters typically do not. Hunters characteristically have lived on the knife edge between starvation and subsistence (more accurately, fasting and feasting). And herein is a lesson of major importance: human-engineered abundance, with artificial storage, on the one hand, versus human-foresworn or -restrained abundance, on the other. The former associated with a population spiral, which remains with us still; the latter characterized by systematic and rigorous methods of population control.

We might now refine all of this a bit further. Earlier on I remarked in passing that the emergence of the neolithic was not an abrupt event, the transition from foraging to systematic farming being, rather, an incremental process in each of its various nurseries around the world. Scholars have puzzled and argued endlessly over the social and cultural nuts and bolts of the shift; my eye, on the other hand, is fixed on more thrilling, albeit ethereal fare—on the metaphysical plane. Still, certain nuts and bolts must be considered if we are going to get the metaphysical business straight.

Consider the following scenario. With the melting of Pleistocene ice, regional climates gradually became warmer and wetter, and edible flora and fauna proliferated (except for the anomalous disappearance of certain big game). It was a potential "free lunch" that some groups simply could not resist, so the theory goes: with food so deliriously abundant, individuals helped themselves to more, and what they could not consume in the short run they stockpiled. Bear in mind that this is not yet farming (plant and animal domestication); it is hunter-gatherers hoarding food (the margin idea) and, ominously, lifting the lid off population controls. Population began to climb, according to this hypothesis, and our hunters slipped from simple foraging into what is commonly called complex foraging, hoarding. In the jargon of population biologists, *resource restraints* were lifted by the invasion of a *new behavioral niche*, and reproductive rates consequently soared. Chances are this burgeoning population is beginning to march to the tune of social classes and hereditary prestige, for so the archaeological record suggests hither and yon in this age of post-Pleistocene largesse.

What happens, however, when the free lunch ends? When, say, the climate turns from salubrious to hot and dry for a century or two, or even more, and the plants and animals one has come to rely on become frightfully scarce? Or there is a sudden influx of immigrants into one's domain, where resources are already stretched thin? Then the awful decision must be made either to go back, as it were, to the simple forager economy, or to press on, through that ominous doorway into outright domestication: the neolithic.

This appears to have been the situation that confronted a number of complex foraging societies worldwide, several thousand years after the end of the Ice Age. A climatic or demographic crisis either drove them back into the secure austerities of simple hunting and gathering or catapulted them into the neolithic, the enchanted realm of sky gods, priest-kings, and a chosen people of history.

My interest, again, lies exclusively in the metaphysical aspects of this process, in the heretical free-lunch response to post-Pleistocene plenitude that in turn set the terms, it would seem, for the even more outlandish (literally: otherworldly) cast of the neolithic mind. But was the free-lunch idea truly heretical? Would it necessarily have been presumptuous for hunter-gatherers to harvest plant and animal persons in greater abundance when these were seemingly in greater evidence? Not necessarily. This could conceivably have fueled (literally) a population bulge, creating a society that would have to make tough decisions when the ecosystem became stressed climatically or demographically, or both.

The real issue, however, is hoarding: the stockpiling of the edible portions of slain plant and animal beings. This is the heresy—a heresy that would be compounded by the neolithic commitment to artificially producing these plants and animals, stripping them of whatever remained of their spiritual volition, their permission—the gift—in the process. In the case of plants, it was probably more a matter of repositioning their spiritual virtue, which I prefer to call their grace, into the theater of the sky gods of weather and time, the keepers of agriculture. Let me acknowledge here that hunter societies have commonly stored certain foods gleaned from the wild, be it land or sea, in historic times. A comparison of the historical record with the more recent ethnographic reports of such storage gives the impression that it was always something engaged in with trepidation, and a sense of risk, while at the same time it is a practice that has gained ground as traditional values have been either compromised or disowned. Even in today's apostatizing climate, the more potent animal beings almost never have their remains dried and put away. In this as in all such complex issues of human behavior, the boundaries are never cleanly drawn, nor are they rigid.

Since we see the roots of the neolithic mentality in complex foraging (the creation of a surplus, social stratification, the beginnings of a population spiral), I shall, for the sake of convenience, from here on include this transitional phase from simple foraging to systematic farming within the metaphysical orbit of the latter, referring to both merely as the neolithic. Hoarding, and a population spiral joined to a profound realignment of social arrangements within the group, seem to lie at the heart of the drift from simple foraging to relentless plant and animal domestication.

We return again to the question, What on earth would possess mesolithic hunter-gatherers to recast their image of themselves vis-à-vis nature in such monstrous terms? Was it a species of greed, perhaps? Greed seems inherent in the free-lunch slogan. I used the term because it is implied, if not frankly expressed, in the models ecologists use in explaining the evolutionary consequences of group selection versus individual selection. More jargon. Group selection operates when all members limit their population (called *fitness*) and, hence, the amount of resources used so that the group as a whole will not overtax its resources and thus may survive. Such a strategy of restraint is always *invasible* by an individual with a strategy of unrestrained reproduction; before the resources for the group give out, this maverick will have contributed a disproportionate number of progeny to the group, merely by reproducing faster. And one can continue playing out the consequences, with the ultimate lesson being (a) the good of the individual might contradict the good of the group, and (b) insofar as the exploitative behavior of this individual is transmitted somehow to his numerically overwhelming offspring, the behavior of the group eventually takes on a similar cast of mind. It is Garrett Hardin's old "tragedy of the commons" idea, where some unscrupulous, selfish pastoralist decides to graze one or two more sheep on the commons than is customary. He gains personally but in the end such selfishness, if widely adopted, results in the destruction of the pasture. And then all suffer.

But I am not moved by the greed argument, and I consider it largely irrelevant, even in today's aggressive and competitive market economy. More fundamentally, I think the underlying motive for the abandonment of simple foraging was fear. As I brood over the metaphysics of the neolithic, I detect a language and artifice ridden by fear: fear of not enough food, fear of animal elusiveness and hostility, fear of our own death. I see none of these elements of fear disclosed in the voluminous literature, both historical and ethnographic, on hunter-gatherers. And that, I think, is revealing.

Inherent in hunter-gatherer belief and performance is the notion that nature is a cornucopia, that nature will always supply one's needs, including sufficient food. Few of us appreciate this about the hunter-gatherer mentality. The popular image, even among many academics (so I have discovered from experience—though ethnologists know better), is of hunter-gatherer desperation over procuring enough to eat. That image is largely a fiction.

It is part of a hunter's sense of caution and respect to speak modestly, even critically, of one's abilities in the chase or even to flatly announce failure. To acknowledge skill and success is tantamount to bragging, which is heard by listening game animals, who would naturally take offense. Time and again Europeans were told by a hunter that he and his little band had failed to secure any game in the previous weeks, when further inquiry, perhaps from another source, revealed that the man was, basically, lying. "It is a general rule among the traders, not to believe the first story of an Indian. He will tell you on arriving that there are no deer, and afterwards acknowledge them to be numerous; that he has been starving, when he has been living in abundance."[8] Was he deceiving to show his reverence to the ubiquitous game spirits and, as well, prompt his white listener to furnish him additional food? Quite possibly: aboriginals the world over have habitually, steadfastly considered Europeans under obligation to furnish them with food, clothing, and hardware quite simply because whites possess these items in seemingly unlimited quantity—"seeing that we are richer than they, [they demand that we] . . . give them liberally whatever we have"—and because whites typically declare themselves to be the aboriginals' friends.[9] Friends, by definition, shared wealth—an aboriginal principle few whites grasped, and, among those few, very rarely embraced with enthusiasm.

The other point is that hunter-gatherers fed themselves differ-ently from, say, Westerners. Hunters both ate different fare and fol-lowed substantially different habits of consumption, all of which has contributed to the fiction of hunter starvation. Hunters are notorious for eating things few Westerners would tolerate today, or in past cen-turies for that matter. Basically, if it yielded calories and could be kept down it was put in the pot. Mammalian innards (including gut contents), bone marrow, and blubber, or boiled whole fish (scales and all), or insect larvae—menus of this sort revolted European guests, who considered these diets marginal, not to mention bar-baric. Nutritionally, however, meals involving these items made good sense. So, too, did boiling one's food rather than roasting or baking it: hunters, historically and ethnographically, tended to con-sume their food as stews, in which the minerals and vitamins were retained rather than being lost in drippings or poured off as waste. Thus we witness a diet European observers oftentimes deemed un-palatable (read, *unacceptable*), while simultaneously we see the con-sumption of an extraordinary range of foods. Our modern diets are impoverished and unimaginative by comparison.

As for eating behavior, here too we must recognize a key dif-ference between Westerners and aborigines. The latter have tended to follow a feast and famine regime well suited, it turns out, to their physiology. Well suited, as well, to the way nature presents herself seasonally and regionally for consumption. This is not the place to discuss hard-core human physiology, except to say that basically hunters practice lipid-loading (fat-loading, analogous to carbohydrate-loading among marathon runners before the big race) when animal flesh is abundant and in anticipation of lean weeks ahead (say, the transition from fall to winter, and winter to spring, when ice freeze-up or break-up makes mobility difficult for more northerly bands). Their bodies respond by slowly parceling out these fat reserves as energy over the days and weeks that follow. ("They diet themselves very thoroughly afterwards, living sometimes for about a week more or less on tobacco-smoke, and not returning to the hunt until they begin to be hungry.")[10]

All of this I have developed in some detail elsewhere; what I wish to convey here is that it was Europeans who found themselves incapable of reproducing this metabolic feat. In short, hunting peoples typically are genetically predisposed to be efficient food storers (chiefly as fat); binge eating ("they stretch their bellies marvellously") followed by protracted food deprivation should not be confused with gluttony followed by tragedy, the erroneous conclusion drawn by many a white observer over the centuries.[11] It is, rather, the way game presents itself in certain seasons of the year—a natural economy of nature to which the hunter's endocrine system is well adapted. Europeans living or traveling with aborigines on these occasions generally complained of starving—true enough for them but not for their companions, who were fasting. More to the point, their hosts were not complaining about the matter ("they can . . . abstain from food for many days together with a degree of indifference and good humour incredible to an European"); they still considered nature to be generous and forthcoming in satisfying their needs—needs, not necessarily desires.[12] There's a difference, and hunter-gatherers acknowledge and accept it.

In this century, as ethnologists have attached themselves to hunting bands, they have frequently remarked on the natives' seemingly irrational confidence in the hunt, in nature's care. The amusing thing is that it was the ethnologists who were worried about possible failure, not the hunters. As we have seen elsewhere, game animals are not imagined as either elusive or hostile. They are more properly conceived and approached as friends who willingly offer their fleshy coats to *Homo*, their ancient kinsman, as the latter has need of them. On the other hand, where hunter societies became compromised by European colonization or trade, we find this kind of confidence in short supply, or even flatly rejected, in favor of reliance on what one might term metropolitan sources of food and other necessaries.

The other key fear that underlies the neolithic mentality, as I see it, is of death. One's own mortality. Here, once again, hunter societies stand in sharp contrast to our own and those of our neolithic ancestors. Again I am assuming that the equanimity that historic and modern hunters have shown when dying is but a continuation of an ancient, I suppose essentially paleolithic, trait. Just as whites have marveled at hunters' ability to withstand prolonged food deprivation, so have they marveled at the ease with which hunters have left this world—"with as little struggle as they come into it."[13] To put it succinctly, it was all right to die, as long as one died naturally, on one's deathbed. It was not so fine to be killed in battle by an enemy from another tribe or band, and it was positively unacceptable to be felled by sorcery or murder from within or outside the group. Traditionally, one's surviving kinsmen sought revenge for unnatural death; the blood of the deceased was thought to cry out for it. Another intolerable kind of death was that which came from mysterious epidemic disease, like that introduced by metropolitan, civilized societies, with their penchant for animal domestication, urban congestion, and extensive trade networks—each an epidemiological disaster. Disease of this nature indicated something was dreadfully wrong with the cosmos, or that massive necromancy was abroad.

Hunting-gathering peoples did not fret about an afterlife until Europeans brought them lurid tales of heaven and hell. (Even then, we are amused to find one theologically askew Indian in seventeenth-century New France inquiring whether the Frenchman's heaven served those divine French pastries.) For hunting peoples, attention was concentrated on this world and this life, and an acceptable part of life was growing old and feeble and dying. Nature, the universe, was not damned for this economy of dying—not in speech or, implicitly, in artifice (there were no elaborate efforts at preserving the human body from decay, for example, as with mummification in ancient Egypt and elsewhere). In death, the spirit decamped and functioned someplace else on the earth; the important point is that there was no anxious speculation about the nature of this place of the dead. It was considered natural, and fine, was not to be feared, and in fact resembled the world and called forth the same mundane tasks which people knew while in the flesh, which is why the dead were buried with essential tools, jewelry, and clothing— grave goods, ethnologists call them. My impression is that grave goods became much more common following contact with Europeans, say, or other exotic traders, perhaps because the spirit place was thought to lack these items—an interesting and provocative thought. Be that as it may, hunters never set for themselves the task of severing the circle of time, so palpably evident throughout creation, to fashion it instead into a vector that might hurl mankind out of earthly creation. Death was but a shedding of the fleshy apparel, a concept and phenomenon already familiar from the hunt. The spirit took up residence where all human spirits resided on earth, as one part of that larger, teeming commonwealth of spirit beings that composed the essence, the chief reality of the earth.

What I am driving at is that neolithic societies had the temerity and, I suppose, the courage to forge into being, orally and manually, a seemingly unprecedented relationship with nature that was the product of some lesion of the imagination. As we reflect on that lesion, let it be asserted once again that there is no convincing archaeological, historical, or ethnographic evidence suggesting that preneolithic *Homo* was somehow constitutionally driven to overwhelm and conquer nature (Pleistocene overkill would not contradict this, incidentally), where the only thing lacking was the creative genius to figure out how precisely this might be technologically accomplished. That proposition, or assumption, which has enjoyed a remarkable currency throughout Western history, may ring true theologically, but it is archaeological, historical, and ethnological fantasy. It is time to thoroughly disabuse ourselves of it. The neolithic, I would argue, was not a technological coup, a technological breakthrough, so much as it was a breakthrough, more in the sense of rupture, of the human imagination. From here on, human oral and manual discourse with nature would begin to take on an entirely new cast, with both of these neuroanatomical powers now enlisted in the service of establishing *Homo*'s primacy and mastery over nature. The essence of the neolithic is cognitive, a revolution in perception and imagination—this is what I am suggesting. The mind that had fastened a check on *Homo*'s rendering of nature in speech and artifice now effectively removed it. Behold the narrative of mankind for the past seven thousand or so years, that is, since the dawn of Mesopotamian, Southeast Asian, Mesoamerican, and other civilizations, as one of exponentially increasing technological and oral license.

Somewhere in the garden of the neolithic, alongside the tiny, delicate ears of primitive maize or the earliest domesticated Old World grains, rices, and legumes, we ought to find as well the seeds of this tumorous unrestraint. I flatter myself that I have: it was when trust in nature turned to distrust, when *Homo* succumbed to the terror that nature would not furnish enough and that we, alas, would die through her malice or alien ways.

# 3

## "Time was the god"

∿∿∿∿∿∿

### The Invention of History

Farming is essentially scheduling. Timing. On the other hand, so too is hunting and gathering. It is a commonplace that hunters are keenly aware of the seasonality of animal and plant abundance and join this to a precise knowledge of the location of the event. Hunters carry elaborate mind-maps, often assisted by some sort of icono-graphical device, of their familiar territory; and they learn from elders and their own experience when to be at a certain spot to en-counter game, plants, fish, or what have you, harvesting each as it appears. All of which is transacted within the perceptual scheme al-ready described, but also within a perception of limit—*human* limit: that humans must restrict their impact on the earth. A principle that hunters expressed in two ways: a very real and functioning human population limit joined to an equally rigorous practice of partial har-vest. There was a ceiling beyond which the land and sea were not to be taken; to do so would be immoral, unnatural, and dangerous— curiously, all of them concepts that neolithic societies rejected, or conceptually inverted. Notice that nature was not regarded by hunters as either parsimonious or exhausted; it was more a case of nature being viewed as forever prolific and giving, where it behooved man and woman to comprehend and accept her schedule and place/spacing of such largesse.

Hunters in a sense were taught from generation to generation to read nature's time instrument—better yet, her scheduling-spacing instrument. Inherent in this time-place perception of plants and animals was the understanding that nature, herself, would determine the terms of abundance, location, and timing of each of these articles (beings, really) of human need. Actually, this is not quite accurate: it was not nature—some monolithic, nebulous entity—that decided; it was the plant or animal beings themselves that decided. Hunters cultivated a relationship with individual plants and animals, not with *nature*—an important point.

∿∿∿∿∿∿∿∿∿

The Seneca, westernmost of the Six Nations Iroquois, have a wonderful story illustrating this. The story is called "The Old Man and the Little Man" and was collected on the Cattaraugus Reserve in the 1880s. It was related at the time in Seneca; what follows is an imperfect English translation.

The tale begins with a man who "lived in a solitary place all alone by himself and he was well off but lived on dry barren land, there was no water, though he was very fond of water." With this pregnant introduction it goes on to tell of his success in locating a colossal underground spring, whereupon he built a dam to create a pond to husband his newfound treasure. Thus established, the man married, raised children, and in due course became the patriarch of a large and prosperous nation of similar dam-builders.

One day a "little, small man" appeared, hoping to reproduce the success of the Old Man and his tribe. Though granted permission to establish themselves on the Old Man's waterways, the little people were too puny to erect dams. So they settled for living in burrows along the banks of the streams, and with time they, too, flourished.

And by now it has dawned on the listener that this is a tale about the beaver nation (the Old Man's people) and the muskrat nation (the Little Man's people)—a cautionary tale about the swimmer people (my term).

Disaster struck when "enemies" (humans) began a series of "unprovoked attacks," first against the beaver people and then against the muskrat people. Under repeated assault, *Castor* (beaver) and *Ondatra* (muskrat) were nearly exterminated. Ah, but then the human persons began suffering grievously themselves: mysterious diseases now stalked them likewise without quarter. While on his way home from such a raid, the leader of one of the war parties had a dream in which it was revealed that the contagion had originated in the wrath of the Old Man, who, the dream visitor went on to disclose, was an exceedingly powerful being. So apprised, the human people now earnestly sought to make amends with the beaver and muskrat people, visiting the desolate ponds to deliver speeches of apology and peace, and offer tobacco.

The conversion from slaughterer to supplicant is exceedingly important. See how it involved the power inherent in words—words the swimmer people understood—and the healing grace of a plant. The new relationship became one of communion through words of apology and peace, as well as through the restorative and narcotic powers of tobacco. Dream messages, incidentally, are very traditional among the Six Nations.

The story concludes by saying that the Old Man now presented himself to one of the human people in yet another dream, declaring "that he was not angry with him because he had killed so many people of his nation but for wasting our bones and wasting our flesh. We are willing that you should come occasionally and take as many of our number as you need, but no more. If you will do this we will hereafter be friends and not enemies. I have heard all your speeches and seen you burn tobacco that sickness might [go] from your people. We will take sickness from you if you will do all I have directed."

And so it was that the five Iroquois nations (later to become six)
forged a covenant of peace and amity with the beaver and muskrat
people, which both sides, human and swimmer people, maintained
inviolate, "until other disruptions complicated the world."[1]

∿∿∿∿∿∿∿∿∿∿

Under the neolithic impulse, mankind's sense of space-time changed
dramatically. Prerequisite to that change, however, was a fundamental shift in the locus of plant and animal volition, plant and animal
*permission*. For the coming of the neolithic marks the onset of the
dangerous process of denying will to individual plants and animals
and repositioning their will in cosmic or meteorological forces. In
practical terms, the scheduling of plant and animal florescence and
reproduction were denied those beings and placed, instead, within
the increasingly remote and esoteric realm of sky beings.

   As hunters had sought to influence individual plant and animal
behavior through ritual and magic, so nascent farmers exerted themselves to influence the great cosmic forces, the sky beings soon
denominated gods, who controlled weather (rain, drought, frost,
warmth, etc.) and the punctuality of the seasons. Priests would
emerge as a special rank of individual uniquely qualified in the art of
cosmic divination and, more subtly and dangerously, time itself, on
behalf of a society now increasingly committed to engineering food
into existence. When to clear the land of brush, when to plant and
when to harvest, when to let the arable rest, when to turn on and
turn off the various irrigation mechanisms (in arid regions), calculation of the number of harvests one might extract from fields per
season and the time staggering of crops per growing season—these
matters of scheduling and spacing, along with the need for group
coordination and control, were all now placed in the hands and rhetoric and imagination of a priestly class swiftly becoming ensconced
as the ruling elite. People now turned to the heavens for guidance in
these affairs—guidance, authority, and power.

For the secret to the riddle of plant and even animal production and reproduction was timing: time, which became a clock. If *Homo* could but contrive a clock, he would stand in a position to control the seminal powers of creation and procreation. A careful watch of the heavens, of constellations and planets in their orderly march across the night sky, of the solstices and the phases of our companion moon, revealed such a gigantic mechanism. Perhaps there was as well the palpable lesson of the lunar–female menstrual cycle, attendant upon woman's possessing the earth's mysterious powers of fertility, of production. If these celestial bodies, so measured in their movements, were divine, or controlled by the divine, then man himself (the masculine gender, here, probably being fully justified) could make contact, and perhaps exert influence. In any case, it must have become clear that the mind of these celestials was mathematically precise and beautiful, and computable by those who made it their business to attend to such things: priest-astronomers. Here, behind the caprice, the willfulness of plant and animal powers of scheduling, of *economy*—of estrus, rutting, and the flowering of plants—lay the higher power, in truth the laws, of celestial time. Quantifiable time. The priests who figured this out must have sensed themselves ushered into the presence of the almighty beings. It was verily a "fall out of nature into knowledge," maybe the Fall that damns us all: bewitched by time. "Things," observed Eiseley, quoting another, "were to grow incalculable by being calculated. Man's powers were finite; the forces he had released in nature recognized no such limitations. They were the irrevocable monsters conjured up by a completely amateur sorcerer."[2] Time's arrow—the cosmic monster that has proved itself infinitely more terrifying than any projectile ever devised by the paleolithic imagination.

∿∿∿∿∿∿∿∿∿∿

To mint animals for *human* time (read *economy*) required fencing and tethering, fodder and shelter and protection against predation, very likely the detention of the newborn as a means of controlling the mother, clocking estrus, and selective breeding for desirable traits through manipulation of the mating process. One can quite properly label the process detention, forced breeding, and patronization (fodder, shelter, and protection)—these constituting, of course, the precise terms of slavery, human slavery, down through the millennia since the neolithic. Small wonder that many scholars detect in animal domestication the ideological origins of human bondage, especially when the human victims could likewise be depersonalized and rendered as bestial, that is, inserted into the below-threshold category of animals, who long before had been imagined and spoken out of the realm of personhood (having volition and giving permission) by that clever speaker and artificer of creation.

Farming societies had a different fate in store for the animals that remained: these became either deified (ancient Egypt and Mesoamerica provide two stunning illustrations of this principle, but one finds it in many if not all other civilizations) or vilified as wild, meaning dangerous, coldly brutal, and withal outlaw. Whether enslaved, enshrined, or outlawed, beasts were still, I maintain, domesticated to human terms. Perhaps the most interesting common denominator here is that each category, in its own way, was victim of human fears.

Deified creatures represent the neolithic fear of the otherwise nebulous spirit realm and the effort to comprehend and, more, emasculate those terrible powers through the creation of animalized deities. Essentially, it was a matter of controlling dangerous spirituality by imagining animalized gods who behaved the way humans did. The result, more accurately, was the domestication of the fear of the numinous, whereas with barnyard domestication, one might say, mankind was endeavoring to domesticate its fear of famine.

It is interesting to contemplate the god of the Old Testament in this context. Jehovah was tireless in reminding his people that he beggared human conception and human likeness, including those occasions when the latter lay thinly disguised in animal form. Indeed, nothing infuriated the Almighty more than the spectacle of the Israelites bending the knee to graven animal images. Through his prophets he pronounced them frauds, raining down terror to drive the lesson home. He was right: these were nothing more than clever impersonations of neolithic man and woman themselves. Which of course leaves the question: Who, then, was this Jehovah? In my opinion, a frank and virulently potent icon of a newly emergent historical consciousness; essentially, a spokesman for the conviction held by a nucleus of individuals that the destiny of the Children of Israel should not be frittered away in parochial issues, such as those which consumed the energies of the animal idols one witnesses in profusion among the pagan societies throughout the ancient Middle East.

I will say more about the postneolithic image of plants and animals in a later chapter; what interests me more in this present one is the dawning of a sense of historical awareness. *Homo sapiens* would become historical in outlook and behavior only with the advent of the neolithic; history, I believe, is very definitely an artifact, a fabrication, of the neolithic—in my mind, the most interesting and by far most important contrivance of that paradigm. Before that, if we can extrapolate from more recent hunter-gatherer evidence, humanity was naive of such a concept; there was no need for history in the world as perceived by hunters. But farmers are a different story. Casting an eye around the world at the various foci of the neolithic, one sees repeatedly the irresistible urge to commit oneself to the siren song of history: the intoxicating belief that mankind is in truth a special creation, a superior being, superior to everything save the unseen and superordinary powers of the heavens, who have chosen this human vessel as their special agent in carrying out their will. The link between sky gods and man was initially forged, let me reiterate, in the repositioning of productive and procreative power from earthly plant and animal beings, possibly even including women, to more omnipotent, more generalized cosmic forces.

Somehow, early in the relationship with these new sky gods, the mediating priest-kings seemingly assumed a sense of personal mission: to carve a divinely sanctioned order out of what they believed to be chaos all about them. Invariably we witness these budding civilizations, or city-states, launching themselves on ambitious building programs, constructing spectacular, even flamboyant, typically colossal temples, pyramids, and plazas, for the glory and pleasure of the heavenly pantheon with whom these potent priest-kings conferred and consorted. Such structures were built as well, no doubt, to overawe the populace. Along with this construction went the amassing of wealth—precious metals, stones, and fabrics worked by armies of artisans into refined objects; frescoes and murals lavished on building exteriors and interiors; statues, idols, and altars; etc.— all this, too, for the delectation of the heavenly pantheon, whose ranks the reigning monarch, his family, and ancestors now solemnly joined. Trade and commerce, like tendrils reaching out and wrapping themselves around the sources of this wealth, would start out as another royal, and divine, prerogative. As would war. Holy wars pushed back the bounds of this celestial center, bringing both benighted landscape and its peoples under the care and fatherly guidance of the Anointed One. Thus foreign lands, or cities, were assailed, to envelop them within the spiritual economy of this, the only legitimate state of humanness. Success in all of these endeavors was undoubtedly taken as a sign of godly favor and alliance. Meanwhile, the corn- or grain-fed population grew by leaps and bounds, becoming the multitude. And thus, one speculates, was history launched in more than one neolithic civilization. History, bear in mind, was initially royal history, dynastic history: the story, more, the recording, of a single ruler or his family, generation to generation, and their divine mission—order out of chaos.

Europe, seething with popular resentment against its own decadent monarchies, would have a glimpse of one such paradigmatic king ruling the Hawaiian islands at the turn of the nineteenth century. Here was a divinity about whom Marshall Sahlins writes: "To

be able to put the society in order, the [new] king must first reproduce an original disorder. Having committed his monstrous acts
against society, proving he is stronger than it, the ruler proceeds to
bring system out of chaos. Recapitulating the initial constitution of
social life, the accession of the king is thus a recreation of the universe. The king makes his advent as a god." He was the god who lived
both "the life of the people" and "the life of the cosmos" simultaneously; the sole individual with the divine right to calculate society's time "in dynastic genealogies, . . . in royal traditions. . . . 'I
was born when Kamehameha conquered O'ahu,' " one of his subjects
might say. Whereas I myself, subjected to not an unlike history, was
born at the end of the Second World War—I might say.[3]

It is worth noticing that the structures erected by these priest-
kings were intended to be permanent—permanent witnesses to a
glorious point in time. The temples, pyramids, palaces, statues,
stelae, and other such monumental artifacts constituted tangible
expressions of history: they left something human, something connected with the activities of a special individual, on the sacred landscape, itself a mirror of the celestial landscape. Even their destruction or dereliction had a history; noiselessly crumbling there in
desert, jungle, or plain, they prompted a certain recollection of human endeavor and ambition. Artifice such as this would inevitably
trigger historical awareness. Here was palpable history, history that
had transformed the land, hard-edged testament to mankind's presence, man's thoughts, his viewpoint—and his valiant separation
from the Surround. Except that the stones, in falling down in bits
and pieces to sand and soil, eloquently suggested otherwise.

In sum, the priest-kings aimed to locate themselves and their
heroics, including their people, in stone monuments, and hence stories, intended to endure. Consider the Egyptian pyramids, Sumerian
temples and palaces, and so forth. The idea was to insert oneself into
this powerful, holy continuum of time, of history. There is something sacrosanct about being able to make one's mark on history

through some monument or achievement. In modern times the effort has been not so much to build a tangible monument as to be remembered by history, to go down in the history books. The point being that with the neolithic and massive, grandiose architectural and engineering projects, the artifacts themselves become historical benchmarks. More, they stimulate a certain kind of story told down through the ages. For they are *there*, and must (it is believed) be explained. (Thoreau, the iconoclast, disagreed, as I do.) Human monuments thus become a means of preempting the stories inherent in the untrammeled landscape, the pristine Surround.

By way of contrast, notice how very different this is from the imprint left by hunter-gatherers, even in recent times. The structures assembled by them are seasonal, impermanent; the landscape reabsorbs (reclaims) them within several seasons of their abandonment. Hunters leave no lasting testament of their passage or presence, reflecting an imagination which eschews such chauvinism. They inflict no disfiguring human history upon the face of the land—its trees, its soils, wildlife, and waters—except, I suppose one might argue, through fire. They consider it existing always as in the time of creation, always original. Nothing is left behind to suggest an alternate, competing story from the one of a cycling, and recycling, land. It is not a trivial point.

Pushing this slightly further, it is curious to observe how civilization denounces hunting cultures for not making permanent structures, and hence indelible changes, on the land, as though hunters are too primitive to manage it. And yet, to build permanently is to sign history—a principle, a metaphysic, that hunters reject. The fundamental difference is not manual and conceptual capacity and ingenuity; it is more properly a difference in conception of self-nonself.

〰〰〰〰〰〰〰

Yet all of this is strictly theory. My aim is not to recapitulate theory but to look instead for certain key signs of a transformed oral and manual discourse with the earth. With that in view, we might consider an interesting feature common to perhaps all of these emerging neolithic civilizations: the idea that the gods either once and for all, or periodically, created cosmos (order) out of chaos (disorder). The usual scenario begins with a coherent, humanly congenial universe summoned into being by the benevolent creator god, who for some obscure reason now permits this new creation to be tempered (regenerated?) by strife. Thus great good (= cosmos) and great evil (= chaos) become locked in titanic combat. Allied with the forces of good in the worldly, mortal plane of this conflict are the chosen people, hurling their mandate, their trajectory, in truth their history, against the agents of evil, error, and folly. The awesome task of maintaining the integrity of creation becomes the special duty, and indeed raison d'être, of the priest-kings and their heirs. In such fashion is history, which was given birth by the god and charged to the privileged people, played out, chiefly through the vigilance and initiative of superordained leaders who become personally accountable for keeping sacred time moving.

The Judeo-Christian cosmology follows roughly this pattern, where the creator-god, Jehovah, conjured into being the universe and this earth and, at a hallowed spot upon this earth, a paradisiacal home for man and woman. Adam and Eve sinned, however, were expelled from the Garden by the angry and disappointed deity and forbidden reentry, and life thenceforth for all of humankind was pronounced a struggle. When vexed yet again, Jehovah destroyed earthly life by sending a global deluge; only one man, Noah, and his family were deemed virtuous, and they and one pair of each species survived the catastrophe—a cleansing, really. In time the offspring of Abraham, the Children of Israel (Hebrews), were launched on their special history: to establish themselves as a large and prosperous nation in the Promised Land. They were Jehovah's people, and their wars of conquest against resident and neighboring peoples were pronounced holy.

As this sacred chronicle continued to unfold, Christ appeared at a unique, unrepeatable point in time, declaring himself the Son of God, and introducing a new paradigm: salvation through faith in him, as opposed to salvation through good works and animal sacrifice. Jesus Christ achieved this through his own sacrificial death (the voluntary, brutal, and humiliating death of the god, giving this sacred history a new twist). From here on, the symbolic blood sacrifice of animals, under the old dispensation (Old Covenant), was replaced by the once-and-for-all blood sacrifice of the Redeemer, which Christians (as Christ's followers were soon to be called) would celebrate through the rite of Communion. Christ exhorted his followers, "Go ye into all the world, and preach the gospel to every creature" (Mark 16:15): the Good News that to be saved from eternal damnation and present folly and despair one must believe in the cleansing and regenerating power of the Redeemer's death and live one's life now as a disciple. With this, Christ's followers also become a people with a mission: to convert the pagan and the infidel. Having accomplished his divine errand, Christ left this world for his heavenly home, promising to return when the time was ripe to establish his kingdom on earth.

In the 2000 years since that epiphany, Christians have been exemplary as a missionary people: to convert the nonbeliever, wherever he or she may be, so to hasten Christ's return. Studying the Bible, Christians have believed that at some point in the near future the time of human sovereignty will come to a chaotic yet divinely orchestrated end in a series of events known collectively as the Apocalypse, culminating in the dread Battle of Armageddon. Sometime in here Christ is expected to return from the heavens and establish his thousand-year reign of peace: the millennium. When that is done, this world will have served its purpose and God will then destroy it; and with it, history itself will go up in smoke. From then on, all will be eternity; the saved go to be with God in glory, the damned gnash their teeth in the unquenchable furnaces of hell.

That is the broad contour of Christian history. The Jews, of course, dismiss Christ as an imposter and still await their true messiah. There is much that could be said, and has been said, about the development of each of these stages of the Judeo-Christian historical trajectory, but it is not of concern to me here. I am interested only in the large brush strokes, as seen from our vantage point in time, particularly the following key issues: mankind's lot in life being one of constant struggle; Jews and Christians as a people with a mission; history being punctuated by episodes of violent destruction; the death of the god changing the terms of history; the god's blood as symbolic of our salvation; Christians reminding themselves of their redemption and mission by drinking the god's blood and eating his flesh (Holy Communion); Christians hastening the end through their proselytization, so to end human history and inaugurate eternity; mankind influencing time, in this case collapsing it; Christians as conquest societies with a noble goal.

Around the turn of the nineteenth century there was added to this sacred view of history a secular, scientific model with some interesting parallels and divergences. I refer to nineteenth-century Western civilization, of course, and I am thinking of the evolutionary view of Darwin and Wallace and their immediate antecedents. The evolutionists saw the development of biological life as a linear, branching process, species always seeking to adapt themselves to their environment, which itself had changed dramatically, often catastrophically, over the many years (several billion, we now know) of its existence. There was thus built into this budding awareness of evolutionary biology a deepening appreciation of geological flux.

Notice that under this latter model of history humanity has lost its noble, exalted mission; *Homo sapiens* is regarded more as an animal possessing culture. Under the new, scientific view, we contemplate the possibility of our own species' extinction through some blunder of our own devising or our failure to adapt successfully to some natural disaster quite out of our hands, such as this planet being struck by a huge asteroid, as has happened on a number of occasions in earth's past. Ultimately, we are assured by astrophysicists, our sun-star will burn out: bloat into a red giant (slowly, silently blow-torching the earth's surface to a cinder and effectively extinguishing all biological life) and then collapse into a dwarf star of enormous mass and gravity. This final prediction perhaps gives history (both cosmic history and human history?) a sense of nihilism, of futility, since such cosmic destruction appears inevitable, there being no way mankind can influence its course.

Under the scientific view, the world retains its geologically violent past and projected future, while, moreover, biological life manages to dance its way through all of this change and mayhem, shifting shape and habits to cope with new circumstances. Withal, the human animal, *Homo*, has come to be regarded as but one such plastic creature within this larger kaleidoscope of life, mankind being set apart, perhaps, by a dazzling ability to fabricate what we like to call, again, culture. And yet *Homo*, that artful, adaptable "creature of the magic flight," might snuff itself out through ideological fanaticism armed with technological tinkering.[4] Failing that, the death of our nuclear fireball, the sun, will cook the genome of all earthly biological life for good.

What we are dealing with are calendars of linear reckoning. Time's arrow, time's trajectory, be it sacred time or scientific time, regardless, it is still a matter of a beginning (creation out of chaos or out of perfect void) followed by the unfolding of irretrievable sacred, biological, or geological events along a ribbon of dates, all of it either combusting or ceasing activity at some point in the future: the river of time transporting mankind, and this world, along with it.

Turning our attention once again to the neolithic, why were these people so transfixed by time? I believe it was because of the farming proposition. Yet planting and harvesting and the whole barnyard phenomenon is a cycle, and best timed as a cycle, whereas this breed of time is linear, developmental, and in some senses cumulative. Why lay hands on the cycle and straighten it into a vector? Again, I am not sure, though I am suggesting it had to do with the priest-kings' vision of themselves as the anointed, lineal, dynastic keepers of order—essentially, their fear of what they imagined as chaos.

Be that as it may, there clearly developed in all of this a new speech and artifice of nature: nature was uneconomical, an unruly thing in need of taming, draining, clearing, domestication, husbandry—all in all, rendered into godly-human productivity, godly-human economy. This was the mandate of the priest-kings and their people. The fabulous successes of the neolithic—the engineering of plants and animals, as food, into delirious abundance—must have confirmed this hubris, this *anthropos-logos*. Nature's myriad, random, wheeling economies would be impressed—coordinated, really—to cooperate within a cosmic economy which civilized societies recognized was more properly a struggle, a pilgrimage, with a beginning and ending. Linear time was the story, in numbers, of the sequential working out of the divine plan through human collaboration; the calendars and other monuments of kingly succession signaled the movement of a process higher and more glorious than unassisted nature was capable of transacting. By becoming linear, even if linear within larger cycles or swings, time ceased being at once mundane, remote, farcical, and discordant, to become now godly, human oriented and human encompassed, and purposeful— the principal instrument of calibrating the focal struggle between great good and great evil.

My purpose is to argue that neolithic societies developed a new species of discourse with nature (*on* nature, would be more like it), not just very different from but vigorously opposed to the animalized, confident, strangely fearless oral and manual dialogue typical of hunter societies.

Look at pre-Columbian Mesoamerica for another dramatic illustration of the centrality and power of calendars in a neolithic civilization, in this case a series of them. Few societies have been so seized by time and history as were the Mesoamericans of pre-Hispanic antiquity. Beginning with the Olmec, on the Gulf Coast of Mexico a thousand or so years before the start of the Christian era, and ending with the imperial Aztec, decapitated by European history in the early 1520s, pre-Columbian Mesoamerica presents archaeologists with a wondrous array of intricate calendars counting the cycles of sun, moon, and various planets. It is a Swiss watchmaker's dream: 13 numbers each gear with the 20 sacred signs (divine day names: the Day Lords) to form the 260-day *tzolkin*; the *tzolkin*, strictly a divinatory calendar, in turn rolls by the Solar Year of 18 months of 20 days each plus the 5 unlucky days tacked on at the end. Wheels ticking and clutching, ticking and clutching past each other in the jungles and mountain plateaus of Mexico, Belize, Guatemala, Honduras, and El Salvador. It takes 52 years for the *tzolkin* to complete all possible permutations of the Solar Year, and then the whole thing starts over again. If the gods will it. The Maya, watchers of the sidereal heavens, probably did not invent this 52-year cycle, which runs as a haunting theme throughout Mesoamerican history. Rather, they gave it fine-tuning.

This is obviously cyclical time, connoting cyclical history, a history marked by creation/destruction/creation/destruction . . . The message of the cycle was always of doom. Civilizations, these ancient sages observed, invariably seemed to wax and wane. Perhaps in an effort to subvert that fatalistic carousel, the wise men created the Long Count calendar. Archaeologists conjecture that it originated with the shadowy Olmec; yet, again, it was the Maya who seem to have refined it. The Long Count introduced the concept of linear time and linear history. Here, it seems, was progress; here was the narrative of dynastic achievement chiseled into the stone stelae in sequential, nonrepeating time. The event would never come back again; it was irrevocably gone. The power of the event now reposed in the numbers themselves—time's signature and time's acknowledgment of the episode, left in passing on the eternal monuments.

It is all
a secret of the zeros unfolding.
Behind, nothing,
before, nothing.
Worship it, the zero, and at intervals
erect the road markers
the great stelae
with the graven numbers.[5]

The kings, in their divinity, sought to postpone the inevitability of cosmic chaos, the ultimate, horrible event they feared would return; the Long Count was their ally.

What is especially interesting is how these Mesoamericans injected themselves into the actual machinery of time and history, the two becoming confused with one another in interesting ways. Time, they believed, was a creation of the sky gods, who had conjured this phenomenon through the shedding of their divine substance: their blood. Even so, this divinely created time was not a pendulum of perpetual motion, for time, like everything else in the universe, ran down, always threatening to stop. To maintain it, humans were obliged to feed it their own divine substance. Kings and queens drew their blood in spectacular displays of self-mutilation, and by the time of the Aztecs the temples reeked of the gore of thousands of human victims whose hearts had been cut out by priests and offered to the voracious sun. (Cortés said the capital had the stench of a slaughterhouse.) For the sun, in time, became the personification of time.

Mesoamericans thus became a people with a now familiar mission: to keep time going by gorging it on blood. By the mid-fifteenth century, the human heart sacrifice had become a monumental state task carried out under the banner of the so-called Flowery Wars. In contrast to the Christian West, which tried to ring down the curtain on history by hastening Christ's return through global proselytization, the Aztecs and their intellectual predecessors sought to forestall the inevitable end of all time (the convulsive end of this final Fifth Sun, the Sun of Movement) by surfeiting it on human hearts. Here, in Mesoamerica, while man harvested corn, the gods of farming harvested manly hearts.

The fact is, these ancient Mesoamericans were both historically minded and practitioners of cyclical myth, which presented them with an acute dilemma when Hernán Cortés suddenly made his appearance on the eve of Good Friday (cycling), 1519 (noncycling): Where did he fit? Moctezuma II guessed wrong and lost his life to that miscalculation. The bearded apparition was definitely a man on an aggressive historical trajectory.

Christianity itself has moved to a cyclical as well as historical tempo over the centuries. One easily recognizes the circular nature of the liturgical calendar, and it is well to recall that European settlers of seventeenth-century New England were in the habit of equating their implantation there with the Hebrew conquest of Canaan. America, for them, would be the New Jerusalem. Despite the seemingly never-ending wheels of experience and analogy in the Christian calendar, overshadowing all else was the great, awesome, and inexorable course of sacred history, a conviction of history that defined Western man's speech and artifice of nature. Western man and woman would now pronounce and manipulate nature as the spokesmen and architects of this cosmic vector of sacred time.

Here was the new image: creation as backdrop and setting for sanctified history. And when the deity was finally dismissed from involvement in that history, in the eighteenth and nineteenth centuries, to be replaced by Providence and Destiny, and eventually by human technological ingenuity and the sheer righteousness of civilized institutions—Progress, in other words—the earth continued to be spoken of and handled as raw material for *Homo*'s continuing lofty endeavors. Along the way there were occasions when certain "resources" were believed by many to be more spiritual and aesthetic than mercantile, when the pastoral or wilderness condition was preached as a restorative for civilization-weary society. But even this is organically linked to history; nature "appreciation" originates as an interlude, a relief, from the dominant narrative, one now holding nearly all of humankind prisoner: the Reign of History. Sadly, in recent decades, both the earth and the concept of wildness have become cognitively trapped as "the environment"—yet another thing for *Homo* to conserve and preserve. Protecting the earth from the insults of Progress (history) has become one of the new and highly fashionable projects of the artifice, rhetoric, and mentality of, absurdly enough, history. But can we not see that it is above all ourselves we trap—on what is true self, true humanness—in the fallacious cognitive inversion that the earth is "the environment" for us to take care of?

Over the millennia it was easy to slide from the sacred into the profane, for the real power Western man had over nature was neither his god nor his technology but his imagination of the awful course of history, the same power that drove the ancient Mesoamericans in their blood-soaked civilizations. The image spawned and propelled by fear.

~~~~~~~~~~~~~~~~

There is another story of creation, markedly different from the divine, historically and humanly ordained one just described. It is the mythic view, typical of hunter societies and even marginally agricultural peoples; and its hero is that bizarre individual, the trickster-transformer (or merely trickster). It is instructive to compare the two ideologies for the biases each reveals and implies about nature and humanity's rendering of it in speech and artifice.

In the mythic view, the original act of creation is invariably an incomplete, rough-hewn affair. This first stage of creation varies substantially from one mythology to another. It is sufficient to know that this primordial world presents itself essentially as raw material in need of refinement; it is a world innocent of spiritual and even physical limits of identity: plants, animals, people, and even elemental forces exist as beings capable of metamorphosing into one another's vessel at will.

Matters change, however, with the murky advent of the trickster, who is often the madcap rascal, always incorrigible meddler and interrogator, the essence of buffoonery, standing somewhere between the Marx brothers and some creature out of Lewis Carroll's vivid imagination. Believe it or not, it is this clown, this scatalogically minded, perduring madman (though often destroyed, he invariably manages to piece himself back together again) who will establish the conditions of life to which man and woman must henceforth conform. These include illness, aging, sex, marriage, and death itself, along with all the other various human urges, needs, and limitations, as well as the nature of animals and even such abstractions as irony and paradox. It would be difficult to call this creating cosmos out of chaos; that does not quite capture the nature of what is involved. Likewise, it is important to notice that the trickster is not strictly human; he is as much animal being as human being. Indeed, the confusion of humans with animals, generally human males with animals, is one of the more arresting features of this primal event.

Primal event. Another term for it is *cosmogony*. The mythic mind never imagines this First Time as remote but as continuing even now, ever fresh, ever capable of being experienced in all its crackling power and originality through the medium of ritual. All acts and activities in the cosmogony are considered exemplary models, blueprints for all mankind, and all other kind as well, all being enjoined to live the way these originals did. For that is the way, the only way, of Power. To conform to the original, cosmogonic model amounts to a ritual performance, whose terms, for human beings at any rate, are transmitted through myth—scores, if not hundreds, of them, each a packet of Power. These are the tales of the ancestors, the heroes, the spirit beings.

Often the myths carried a warning, built around the career of the trickster or the exploits of some other powerful individual whose curiosity, greed, pride, or other (human) character flaw got him, or her, flattened—the funny part—while triggering in the system a chain reaction with permanent, and sometimes dire, results—the sobering part. The story might end with the narrator intoning, "And this is how this relationship came into being."

It is interesting how often these tales stress the sanctity of relationship, commending respect for the unseen forces that bond all of the players (animal, plant, human, and elemental) to one another. Thus the myths underscored that humans were not to tamper foolishly with the other sentient beings. At the same time, the audience learned to be wary of any innovation, including technological invention, that did not have a mythic referent; innovators could reap the whirlwind for their temerity—they could upset the forces of proper relationship. (Innovation introduced by powerful alien beings, however, could be a different matter, as contact with Western Europeans would demonstrate.) Generation after generation was reminded that the earth continues to be an exceedingly powerful, prickly place and that the boundaries of human engagement with its errands, not least in human speech and artifice, had been defined by the originals, not a few of whom had suffered grievously for the urge to experiment. Nobody in his right mind would want to upset what had been already established through such exemplary trial and error.

Certainly nothing here suggests a historical trajectory of a chosen people. Time, in mythic thinking, is a very different order of perception, obligation, and experience. Here, time's instruments of measurement are the myriad cycles of nature and its citizens, all thought to be taking place within the suspended moment, the still point, of creation. No image of a linearly receding past, nor of a terminal future. Perhaps the sentiment which kept it all going, and palatable, was that of being taken care of—the absence of fear—the conviction that the other-than-human beings would never cease conversation with the human beings. Neolithic speech would render this formula as: the confidence that nature would never expel man and woman from her garden of delight. But mythic societies would never have thought to put it this way, for there was no entity *nature*, and no *garden* of fruity delectables. That was the dream, the fantasy, of the desiccated, oriental urban or pastoral mind. Rather than a world conceived in violence we perceive here a world conceived more in comedy and the picaresque. Little room for human or divine tragedy here, nor is there provision for divinely sponsored violence in pursuit of noble agendas. Just the matter of muddling through, of surviving by adapting, imitating, and otherwise exercising one's imagination to its fullest and most sensual capacity.

~~~~~~~~~~

*On the 24th day of the 4th month of the year 819, Han Yu, Governor of Chao-zhou (Canton), instructed his officer Qin Ji to take one sheep and one pig and hurl them into the deep waters of the river Wu as an offering of food for the crocodiles. When the crocodiles had gathered, Han Yu addressed them in the following manner:*

"In ancient times, it was the practice of our former Emperors to set the mountains and swamps ablaze, and with nets, ropes, spears, and knives drive beyond the four seas all reptiles, snakes, and malevolent creatures noxious to man. Later, Emperors arose who were of lesser power, unable to maintain an Empire of such vastness. Even the Center was forsaken, let alone here in Chao, between the five peaks and the sea, 10,000 miles from the capital. In that chaos, you crocodiles crept back and multiplied. It was a natural situation under those circumstances.

"Now, however, a true Son of Heaven has ascended the throne: one godlike in wisdom, benevolent in peace, merciless in war. All within the four seas and the six directions is his to rule, administered by governors and prefects whose territories pay tribute to furnish the great sacrifices to Heaven and Earth at the altars of our ancestors and all the gods.

These governors and crocodiles cannot share common ground.

"The governor, under the command of the Son of Heaven, has been entrusted with the protection of this land and of its people. But you, bubble-eyed crocodiles, you are not satisfied with the river depths. You take every opportunity to seize and devour people and their livestock, bears and boars, stags and deer, to extend your bellies and multiply your line. You are thus in discord with the governor, and seemingly rival his authority.

"A governor, no matter how feeble, could never bow his head, humble his heart before a crocodile, nor could he stand by in trepidation, shamed before his officers and subjects, acting in an unworthy manner during the existence granted him in this place. Therefore, having received the command of the Son of Heaven to come here as his deputy, he must contend with you, crocodiles. If you have understanding, hear then the governor's words:

"To the south of this province lies the great sea. In it there are places for creatures as great as the whale or shark, insignificant as the shrimp or crab. All there have a home in which to live and eat. If you left this morning, crocodiles, you would be there tonight. Thus I will make this agreement with you:

"Within three days, you must take your hideous brood and head south to the sea, thereby submitting to this appointed deputy of the Son of Heaven. If three days are insufficient, I will allow five. If five days are insufficient, I will allow seven. If, however, after seven days you still linger, with no indication of departure, I will assume that either you have heard and have refused to obey the words of your governor, or else that you are vacant and without reason, incapable of understanding even when a governor speaks to you.

"Those who defy the deputies of the Son of Heaven, who do not listen to their words or refuse to accept them, who from stupidity or lack of intellect harm people and the lesser creatures—such as these will be put to death. The governor will select skilled officers and men who, with strong bows and poison-tipped arrows, shall summarily end this matter, not ceasing, crocodiles, till you all are slain. I therefore recommend that you do not forestall your decision until it is too late."

*That night, a violent storm struck the province. When it subsided, some days later, it was discovered that the crocodiles were gone. They were not seen again for a hundred years, when the Empire was again in ruin.*[6]

# 4

## "I am inside you, have no fear"

∿∿∿∿∿∿

### The Illusion of Self-Nonself

It has been nearly fifteen years since the South African novelist J. M. Coetzee published his biting satire of conquest and colonization in a short story he called, merely, "The Narrative of Jacobus Coetzee," an account of a Dutch merchant's disastrous elephant-hunting expedition into Hottentot territory in 1760. The tale is completely riveting, distinguished by its ethnographic accuracy and violent and deranged colonialist sentiments, a mockery of all those historical narratives which treat of the reduction of the savage to civility.[1]

The narrator, Jacobus, is a relentless imperialist and frontiersman, who imagines himself plundered and then captured by unassimilated Hottentots he happens to meet along the way. He vows to avenge himself should he providentially gain his release. His captors (or are they hosts?) do indeed insist that he leave, after he has committed what in their eyes is a barbaric act (mutilation of a child who was taunting the pitifully ill Afrikaner). In great satisfaction he quits their squalid encampment, accompanied by just one of his Hottentot servants, the remainder of his native retinue having abandoned their master's brutality for the more congenial company of the villagers.

Jacobus relates his tortured yet successful return to civilization, and the raid he sponsors shortly thereafter. In a scene that could have been lifted from the Old Testament, the village is put to the torch; and one gets the impression that its inhabitants are systematically butchered (men, women, and children), while the Dutchman's erstwhile staff is executed, to his enormous satisfaction and amid his expressions of Christian piety.

There is a point within the tale where our colonial superman delivers himself of an exquisite reverie on the metaphysics of the gun and enumeration. It constitutes, I think, the finest and most incisive statement I shall ever see pinpointing the implications and reverberations of that historical mentality which was conceived in the neolithic.

Jacobus begins by confessing that "in the wild I lose my sense of boundaries. . . . There is nothing from which my eye turns, I am all that I see. . . . What is there that is not me?" The question is absolutely primal, embracing both the biological (being the most fundamental question confronting the body's immune apparatus) and the social realms—to wit, what constitutes self and what nonself? The historical mind responds very differently from the mythic. "I am a transparent sac with a black core full of images and a gun," announces the historically minded Jacobus.

> The gun stands for the hope that there exists that which is other than oneself. . . . The gun is our mediator with the world and therefore our saviour. The tidings of the gun: such-and-such is outside, have no fear. The gun saves us from the fear that all life is within us. It does so by laying at our feet all the evidence we need of a dying and therefore a living world. I move through the wilderness with my gun at the shoulder of my eye and slay elephants, hippopotami, rhinoceres, buffalo, [etc.] . . . ; I leave behind me a mountain of skin, bones, inedible gristle, and excrement. All this is my dispersed pyramid to life. It is my life's

work, my incessant proclamation of the otherness of the dead and therefore the otherness of life.

Self and nonself are sharply delimited here, even adversarial. The rifle assures Coetzee of that illusory separation. Further guarantee is given by his culture's penchant for counting, for quantifying nature. "We cannot count the wild. The wild is one because it is boundless." Once more Coetzee approaches that terrible dilemma of self-nonself. But the neolithic rescues him:

> We can count fig-trees, we can count sheep because the orchard and the farm are bounded. The essence of orchard tree and farm sheep is number. Our commerce with the wild is a tireless enterprise of turning it into orchard and farm. When we cannot fence it and count it we reduce it to number by other means. Every wild creature I kill crosses the boundary between wilderness and number. I have presided over the becoming number of ten thousand creatures. . . . I am a hunter, a domesticator of the wilderness, a hero of enumeration. He who does not understand number does not understand death.

Nor does such a person understand history, I would add.

At the end of this little episode, Jacobus acknowledges that "the need [for the gun] . . . is metaphysical rather than physical. The native tribes have survived without the gun. I too could survive in the wilderness armed with only bow and arrow, did I not fear that so deprived I would perish not of hunger but of the disease of the spirit that drives the caged baboon to evacuate its entrails. . . . Every territory through which I march with my gun becomes a territory cast loose from the past and bound to the future." The trajectory of history.[2]

Here stands *Homo historicus* triumphant, rifle in hand, foot planted on the head of the slain Other. Coetzee's terrible gun has become the perfect instrument of an imagination of fear. Where is the language of courtesy in all of this? Of restraint? Where is the animalness and plantness of this man's speech? By imaging the Surround into the fearsome Other of wilderness, and imaging animal and plant beings as nonself, the culture which bore the Dutchman Jacobus Coetzee reveals its rootedness in fear, fear reaching all the way back to the original domestication of the place and its other-than-human beings—back to the neolithic. It is inevitable that an imagination of fear will create a language and artifice of fear, this being but another way of saying that an image of fear will create a language and artifice of death. Jacobus Coetzee is an eloquent and persuasive spokesman for the impeccable reasonableness of a civilization of death. Note that civilizations of death begin with the micro-surgery of separating self from nonself, with the specious metaphysic that "such-and-such is outside, have no fear." Coetzee enunciated it brilliantly.

Yet, if one is going to disengage from such primal and mythic linkage with the Surround, one must perforce relocate oneself in a different dimension, a different matrix of being and meaning. History, the river of time, I would argue, becomes the perfect—more, the necessary—substance and agency of that speech and artifice which seeks to separate self from nonself. The dynamic of such a mad act of aggression inevitably produces a civilization committed to institutionalized, ideological acts of death. The imagination of history in and of itself seems to me to encourage such mayhem—ironically, always in the name of order—supposedly in the service of cosmos, of law. At its most fundamental level, historical conscious-ness is a death machine; it is a requiem that both sings the Mass and wields the knife—or the rifle.

Starting out as the chronicle of priest-kings, history originally was the oral or literary or visual rendering of that exalted being's successes in combatting the forces of chaos, of dissolution, of evil. The ancient Greeks present something of a special case, for here the state, the polity, in certain cases is the guardian of cosmos. But right now I am more interested in the artifice born of this new, neolithic-inspired historical consciousness, in particular the gun. Other authors have amply charted the historical development of technology in civilizations; it is enough for me to identify the neolithic as the major conceptual, oral, and artifactual watershed separating the hunter-gatherer conception of the world from our own. From the farming view there would in time develop, depending on location, an increasingly urbanized and secularized, and eventually mechanized and industrialized conception of the earth. Rather than trace that process from the dawn of the neolithic to modern times, I prefer to consider those elements of speech and artifice that have emerged victorious or ascendent in our own time. To me, the gun is an especially potent symbol and instrument of modern society's changed image of the earth, albeit just one of many symbols and instruments one might select.

~~~~~~~~~~~~~~~

There is another powerful story, another hunting tale involving a meditation on the metaphysics of the gun, which one might compare to Coetzee's. This one is William Faulkner's "The Bear," part of *Go Down Moses*, wherein a boy, Isaac McCaslin, finds himself joined for two weeks each year to a small group of hunters ritually scoring their virility by sportfully slaughtering deer, black bear, and other creatures in a pristine tract of Mississippi bottomland. Ike McCaslin, like Jacobus Coetzee, comes equipped with a gun, "a new breech-loader, a Christmas gift." With this instrument he has the folly to seek out the magisterial and mythically proportioned Old Ben: a bear who is the spiritual-carnal keeper of the game in this uncivilized landscape. Day after day he looks—expert woodsman, now, with rifle, compass, and watch. He hunts until old Sam Fathers, "son of a negro slave and a Chickasaw chief," stops him cold with the reproach, "You aint looked right yet." Ignoring the boy's protests about his impeccable protocol, the old Indian bores in deeper: " 'It's the gun.' . . . *The gun*, the boy thought. *The gun.* 'You will have to choose,' Sam said."[3]

By this point, Ike burns to have a vision of Ben, not to shoot at him. Nonetheless, the rifle remains an obstacle; the bear refuses to be approached with what Loren Eiseley, recalling an encounter of his own with a fox pup, labeled "upright human arrogance."[4] The rifle— premise and instrument of fear, artifice of self-nonself, pretense of protection—with all of its clinging meaning, is surrendered. Even that is not sufficient. The problem, Ike soon realizes, is that he is attempting to navigate his way to a detached look at the bear using watch and compass, when the truth of the matter is that Old Ben has been tracking him the entire time. How absurd. The bear is demand- ing complete surrender, absolute faith. Utter imagination. At last, "he removed the linked chain of the one and the looped thong of the other from his overalls and hung them on a bush . . . and entered it."[5] And became lost to celestial coordinates and mechanical time. Of course, that is when the bear materializes; the boy gains his vision.

And the message of the bear to the unencumbered boy? Effec- tively, "I am inside you, have no fear." Compare this to Coetzee's tidings of the gun: such-and-such is outside, have no fear. Which interlocutor is correct, one wonders? I would argue that as long as we insist on encountering creation with guns and compasses and watches, along with all the other assorted paraphernalia of fear, we will perforce be persuaded by the logic and metaphysic inherent in these very things. Call it the reign of tools, if you will; Eiseley did. What gave the porpoise its innocence, he realized (in the words of a biographer), was "its freedom from the tyranny of tools and the power they confer on their users," that is, its "inability to manipulate its environment."[6]

Yet the power of tools must start with the power of an image, an image not of the tool itself but of the thing, the being, against whom it is to be directed. The tools of our civilization are artifacts of the neolithic imagination which succumbed to the illusion of self- nonself. I have called it an illusion; I could just as rightly call it a fear:

the most fundamental, truly primordial fear of human existence. Hunter-gatherers confronted and dismissed that fear daily. Farming civilizations and pastoralists on the other hand armed themselves in the death-grip of it. For they had lost, not Eden, as they claimed, but the discipline and ability to imagine, to envision all as self in the vision quest.

ᴧᴧᴧᴧᴧᴧᴧᴧᴧᴧᴧᴧᴧ

Self-nonself: This, I think, has been the critical apprehension of *Homo sapiens'* imagination since the beginning of human consciousness. The mythic mind amalgamates the two in stupendously complex and sophisticated ways. Reading hunter stories and myths, one is immediately struck by the rampant, outrageous shape-shifting. Everything seems alive, sharing a common aliveness and being, and, ideally, capable of slipping into the vessel and role of something else. The trickster-transformer is the supreme embodiment and expression of this strange principle: he investigates, he explores, he lives, he creates, all by the dreamy, childlike powers of transformation and unbounded communion. A vast imagination. We moderns, by comparison, are put off by all of this; we judge it the idle amusement of juvenile minds. Children do indeed resonate to the same stories, the same mentality, one assumes because it is essential and natural to do so. The question becomes, Why have we lost this imaginative dimension as adults, both culturally and, within our own lives, generationally?

The answer, I suggest, has to do with the manner in which we produce food: by agricultural rather than through hunting and gathering means. The answer, I think, is that to produce food on a strictly human-oriented economy, rather than on an economy that is animal and plant oriented, one must first imagine these other beings out of the realm of shared personhood with humankind. All this has been said before. Now, however, we might recast those thoughts a bit and see them as an issue of self-nonself: the domestication of these other beings into an endlessly productive role for human consumption requires an image, then a language and artifice, of nonself. In our own time we have taken the measure of so-called primitive hunter cultures by evaluating their artifactual and technological repertoire. This is an ex post facto measurement, for artifice is born of image, of imagination. The real point of comparison is between hunter consciousness of self-nonself and our own.

If the trickster-transformer serves as the model for members of hunting societies, the vision quest serves as the means for achieving that blending of self and nonself in one's own life. In this manner do males and females in hunting societies invent themselves, truly create themselves: not through exclusively human institutions and referents, not through schooling, as we speak of it, not by asking other humans what it means to be human and humane, but through living, timelessly, with another being that takes genuine care of them. In the Ojibwa story "The Woman Who Married a Beaver," it is emphasized that "never of anything was the woman in want." The language is quaint but the message profound.[7] Moreover, the element of timelessness in the vision is important, and should be contrasted with our own sense of time's fretting, a sensation never present during the course of the vision. Only after learning this fundamental lesson, shattering the superficial illusion of self-nonself, can one launch a career as hunter or gatherer or preparer of (sacred) food. Hunter artifice (technology) and speech follow from this organic image, this mythic constitution. What troubles me most about neolithic societies and their modern heirs is the absence of this critical exercise.

〰〰〰〰〰〰

This seems like a good place to pause and reflect on the knotty problem of taking another's life: the metaphysics of killing. Here we must be extremely careful in the words we use. *Homo sapiens* is a creature of the general category that lives by consuming the tissue of other organisms, animal and plant. This breadth of diet makes *Homo* an omnivore, a trait our kind has housed and honed for eons of evolutionary shape-shifting. The earth seems fond of making things that nourish themselves from the tissues of other beings, living and dead. It would be tedious to try to recount the number and kinds of other animals, besides humans, which slay their food. Yet it is important to keep in mind that we were firmly a part of this tradition long before we were recognizably human.

Is it possible that other carnivores engage in intense moral debate over their consumption of other flesh and live plant matter? The question is, of course, impossible to answer, and, anyhow, seems wrongly put. Perhaps closer to the mark would be to wonder whether other creatures besides ourselves hunt and, even, gather in a manner that might be called courteous or aesthetic. Though maybe this, too, is a hopelessly human way of viewing things; I suspect it is.

What impresses me, from my own background and reading in biology and discussing the issue with biologist friends, is how deeply interdependent biological systems are; predator and prey are stamped with one another's psychic personality, or nature. The one mirrors the other in a vital, vibrant way. Wolf and caribou are joined in the mind, as are wolf and moose; and one sees wolf becoming a hub, or node, around which a number of herbivorous species are joined in a network of the mind and place (habitat). Careful observation of wolf communities suggests that wolves *know* this, and, although causality is impossible to assign, they nonetheless hunt in measured fashion, not recklessly or wastefully; they appear to recognize themselves as a part of the place and prey. This mutuality of embrace—of predator to prey, of prey to predator, and of both to place—is nature's seeming blueprint in these matters. Nothing is psychically abstracted from its habitat and sources of sustenance, except, alarmingly, *Homo sapiens*.

We can discern, here, the absolutely central role of the self-nonself paradox in human awareness. I would maintain that the human equivalent of the wolf-caribou, wolf-moose, wolf-place relationship is the hunter's conviction that all *out there* is self. To kill for survival becomes a transformation rather than a murder. Certifying it as a transformation is the fact that it is eaten—very important. In this manner, moose becomes me, and I become moose, and so forth with every edible creature I snare or trap or spear or net or gather from this teeming landscape where my imagination and powers are embedded. This, I feel, is the best way to view the metaphysics of the hunting-and-gathering proposition. The meal is sacred; it is mythic (the First Meal, while reminding me of my everlasting kinship with these beings); it is the bringing forth of new life, that of the communicant becoming now, in the process, a part of the creature (plant or animal) consumed. It is not insignificant that women tend to prepare the meal of flesh and vegetable in hunter societies, for mythology designates woman as the originator of new life. Eating confirms my selfness with what I consume. It is rather like the transformation that occurs during the masked dance: the adorned dancer has actually become the creature that is being depicted in mime and likeness; the dancer is transformed, translated, because this is the law of the earth: Life is an endless process of transformation/translation.

Hunting peoples stress courtesy and restraint in the hunting and gathering process, because they know, as wolf knows, that at the very core of their being they are in fact these creatures whom they consume. Through their speech and artifice they translate these creatures into themselves, and vice versa. They can do so only because they *are* these beings to begin with; mythology reminds them of this time and again. Thus, I must hunt walrus with courtesy, with restraint, and with the animal being's permission, because I *am* walrus. If I consume it, I am it. The moral dilemma comes only when the communion of interpenetration is cut, when selfness becomes cast in different terms within a different constellation of references. This is what seemingly transpired during the neolithic. It was not just that animals and plants were deprived of their volition; man and woman began turning away from actually translating themselves into these other beings, with consequences that can only be called disastrous—for all involved.

〰〰〰〰〰〰〰〰

In the end, Faulkner's magnificent bear is killed, yes, though it is a death in which he participates. The integrity, the necessity of wildness seem no longer understood by humans, who with their "pretty toys" (Thoreau) keep gnawing away at the edges of Faulkner's great tract of "brooding and inattentive wilderness" to transform it into marketable timber. There is no place, no time—but mostly, no language—for Old Ben's time: mythic time. Sam Fathers realizes this; in the throes of his own death (he "just quit" when Ben was sacrificed), Sam speaks once more in his ancient, native tongue, completely unintelligible to those about him.[8] But Old Ben seems to insist that he be dispatched with a remnant of courtesy, and in this, again, he has his way. While the mongrel dog, Lion, holds Ben by the throat, the equally mongrelized Boon Hogganbeck hurls himself at the towering, rampant creature and plunges the knife.

Faulkner's story is astonishingly true to ethnographic detail on bear hunting among Canadian subarctic Indians. My hunch is that he consulted, among other sources, A. I. Hallowell's published doctoral dissertation on "Bear Ceremonialism in the Northern Hemisphere" (1926) before writing his own fictionalized version of such a hunt.[9] As Hallowell discovered, and numerous ethnographers have since verified, bears are regarded by subarctic hunter-gatherers as being paramount over other creatures in spiritual potency. Faulkner gives this knowledge to old Sam Fathers, the mixed-blood Chickasaw who instructs Isaac McCaslin in the rudiments of the hunt. Bears were, in fact, venerated throughout the continent aboriginally, not just in the Canadian north; it is not inappropriate for Faulkner to assign these sentiments to a southern tribesman.

Of all the creatures in the forest or bush, bears were traditionally thought by North American Indians to most resemble human beings in temperament, omnivorousness, sociology, even appearance when skinned. Yet bears were considered much more spiritually powerful than humans. Much wiser. Like other spiritually charged beings, bears were believed to hear what was being said of them, no matter how remote the conversation. They lived mostly as invisible spirit beings, who from time to time dressed up in fleshy clothing, as a man would pull on an overcoat, and presented themselves to humans to serve as food and raiment. They are thus friends of humans, elder brethren, if you will, who pity us and take care of our needs—as long as they know they are being treated respectfully.

Respect involves a number of things, among them a language and artifice of courtesy. Notice how mankind's relationship with the bear does not begin with the burning necessity of the hunt. Traditionally, there was no such sense of urgency. The food and other tangible gifts that hunter societies acquire from animal beings are but peripheral expressions of a deeper relationship and need with and for these elder beings. Perhaps the best way to put it is that bears were *good to think*. More than good to think: necessary to think. Necessary to imagine.

As far back as human memory can reach, bears have been humanity's most potent teachers in the art of living and dying and being reborn as the resurrected one. In short, bears have always been the most eloquent and forceful witness to the cosmic gospel that we, *Homo sapiens*, are being taken care of. In all of our needs, in life as in death—taken care of. The bear illustrates this principle by deliberately burying itself in the ground, in the hollow of a tree, or the recesses of a cave, for months on end, asleep in metaphorical death, devoid of food and water, dreamily licking its forepaws, alone. And in the spring it resurges to life: healthy, well-muscled, and, if female, perhaps nuzzling her offspring, born and nursed in that wintery tomb.

The bear taught us valuable dietary habits, too, by illustrating the astonishing range of foods that its human brethren might also eat and, equally important, the physiological art of feasting and fasting. And withal, the bear surrendered its own flesh to the attentive hunter and his family, especially at those transitional times of the year when other game had either migrated or were difficult of access. Bear flesh, sacred flesh, is transformed into sacred food (medicine) by woman; hence woman's power of life is now joined to bear's power of life to nourish the household in a feast of communion, sacred food to be consumed all at once, out of courtesy to great power. When all is done, the speaking drum is handed around and played and a song is sung by the hunters, the song of appreciation to a grandfather or grandmother. Bears, hunters knew, were good to sing.

And the message of the bear in all of this? "I am inside you, have no fear." No fear indeed. To encounter a bear in the bush was a blessing rather than a terror; one addressed such an elder brother (sister) as *Grandfather/mother, Old One, Dark One,* or some other term of respect and even affection. But there was no fear involved. Even in rousing the bear from its den, the careful hunter would make an impassioned speech of apology, explaining his family's need and appreciation. The grandmother would then be killed, ideally by the man's own hand, rather than through the agency, in modern times, of rifle—considered cowardly and discourteous. Maulings are extremely rare, so ethnologists tell us. Once down, the slain bear is addressed once more, with further outpouring of heart-felt regret and thanks. And the spirit of the slain one watches and listens, following along as the body is dragged back to the campsite, to ensure that etiquette is observed and to savor the wonderful aroma of the tobacco the hunter sprinkles in the fire before initiating the feast. And, of course, to enjoy the songs.

One *thinks* bear first, and in imaging and mimicking it, one becomes bear, as it were. Only after becoming bearlike is one in a position to consume that flesh. The image is first, the oral and mechanical artifice second, reflecting that image, and the meal last—as symbol, as consummation, as affirmation.

∿∿∿∿∿∿∿

Even now, at the end of a bear feast, James Bay Cree pass the large, discoid drum around the lodge and sing to the spirit of the slain kinsman. The living drum sings, too. The animal teaches its song in the vision or the dream; plants likewise teach theirs, either in the vision or in some other power experience.

In *Good, Wild, Sacred*, the poet Gary Snyder has described his travels with Pintubi Aborigines near Alice Springs, Australia. As their truck thumped along an ancient dirt track his host, Jimmy Tjungurrayi, began rapidly "talking about a mountain over there, telling me a story about some wallabies that came to that mountain in the dreamtime and got into some kind of mischief there with some lizard girls." Then quickly on to another story, about the next hill they had come upon. And so on. "I couldn't keep up," says Snyder, who soon realized "that these were tales to be told while *walking*, and that I was experiencing a speeded-up version of what might be leisurely told over several days of foot-travel."

Snyder later discovered that these stories of place were really *songs* of place, a kind of geography or cartography through singing. Song lines. He relates a wonderful anecdote about camping one night near the Ilpili waterhole with a small group of Pintubi:

> Through the night, until one or two in the morning, Jimmy Tjungurrayi and the other old men sat and sang a cycle of journey songs, walking through a space of desert in imagination and song. They stopped between songs and would hum a phrase or two and then would argue a bit about the words and then would start again, and someone would defer to another person and would let him start. Jimmy explained to me that they have so many cycles of journey songs they can't quite remember them all, and that they have to be constantly rehearsing them.

Next morning Snyder jokingly asked Jimmy, "Well, how far did you get last night?" "Two-thirds of the way to Darwin," came the half-serious reply. For these people, sacred place is conveyed most pungently through song. In a very real sense they knew where they were through map singing, not map drawing, with its unmelodious lines and coordinates.[10]

Voice: the key being that it is the voice of the place-beings themselves. Ruth Murray Underhill illustrates the point in her *Singing for Power: The Song Magic of the Papago Indians of Southern Arizona*:

> A man who desired a song did not put his mind on words and tunes: he put it on pleasing the supernaturals. He must be a good hunter or a good warrior. Perhaps they would "like his ways" and one day, in a natural sleep, he would hear singing. So does the Papago interpret the trancelike state of the artist who derives his material from the unconscious. "He hears a song and he knows it is the hawk singing to him or the great white birds that fly from the ocean." Perhaps the clouds sing, or the wind, or the feathery red rain spider, swinging on its invisible rope. . . . The Papago sternly hold to the belief that visions do not come to the unworthy. But to the worthy man who shows himself humble there comes a dream. And a dream always contains a song.[11]

A dream song, taught by the powers of the place. Humans might learn these songs, so aboriginal societies discerned. Yet there is more involved here: by learning where one is through the song taught by the place, one discovers likewise *who* one is. Hence, who one is is determined and transmitted by the powers of where one is, and that knowledge of where comes, significantly, from the music inherent in the place-beings themselves. Again, we humans have the capacity— and obligation?—to tune our imaginations to the songs of place, and so fix our bearings on where, and who, we are. I said earlier that hunter-gatherers viewed nature fundamentally as song. In Snyder's words, the earth is essentially a "place of singing and practice, a place of dreaming." Indeed, Snyder goes so far as to trace the origin of poetics to the primal "sense of the universe as fundamentally sound and song."[12]

Several summers ago my seventeen-year-old daughter, Lindsey, flew out to New Mexico to join a group of a dozen or so other kids her age, plus two or three counselors, trekking through the Four Corners area: sleeping out under the stars, cooking over piñon pine fires, climbing through Anasazi ruins, through lava tubes, scrabbling up mountains and down into canyons. Very sensory, sensual, and integumentary. It was a powerful experience for her; she gained an identity from all of that place that she now cherishes more than anything else in her possession. For a kid from Princeton, New Jersey, it was something of a vision quest. On her return, her stepmom and I pressed her to tell us about it: the stories. And she obliged. Wonderful stories, told with consummate skill.

But I have noticed something curious. I have seen that when Lindsey wants to recreate the power and essentialness of that experience—the where that has defined the cherished who of her being—she sings the songs she learned around the campfire. Those songs (to me, just the usual campfire songs) were taught by the place, she would maintain. She learned a great many, and as she sings them to herself she sings the experience and, more, the place to which she is now joined and, equally, the who to which she is now attached. I no longer ask her to tell me the stories of New Mexico; I ask her to sing the songs. Like Jimmy Tjungurrayi, she conjures herself into that place and moves through it in song.

Words. "A word has power in and of itself," writes N. Scott Momaday in a meditation on his people, the Kiowa (who "had conceived a good idea of themselves; they had dared to imagine and determine who they were"). A word, he says, "comes from nothing into sound and meaning; it gives origin to all things. By means of words can a man deal with the world on equal terms. And the word is sacred."[13] Sacred, yes, words can be sacred. But they can also be dangerous, when *Homo sapiens* understands and knows only his own words, not those of other place-beings. On the other hand, we have been longing for the wrong sort of communion with other creatures;

other place-beings may not express themselves in words at all, but in song.

It is indisputable that all of nature expresses itself in sound: wind, storm, rain, the sea, rivers, and lakes, volcanoes, together with all the creatures of the air, land, and oceans. True, not all of this sound is within the range of human hearing. When bats, porpoises, insects, and shrews, to name but a few, climb several octaves in their singing, into the ultrasonic range, we lose perceptible reception, although we may receive their singing in less conscious ways. At the other end of the spectrum, scientists have in the last several decades picked up the ocean-spanning deep-sea music of humpback, finback, and blue whales. While, more recently, biologists studying elephant communication in national parks in southwestern Africa have recorded a remarkable repertoire of songs, sung mostly by females. These songs, like those of whales, lie in the low-frequency, infrasonic range. Humans seem able to detect this infrasonic proboscidian singing as a low rumble.

Singing elephants, singing whales: a singing earth. With our formidable imagination we can comprehend these sounds as music, as song. Elementally powerful music. The best, the truest music. Music that charts for us where we are, and following on that, who we are. More, humankind can reproduce that music, thanks to the neuro-anatomical powers of sensory-motor cortex, tongue, lips, and larynx. We can also mimic it by instruments fashioned by those wonderfully dexterous hands of ours, through our stupendous powers of artistry. Music which a singing world can then hear. Man the orator, man the artificer. This might be more truthfully, more provocatively rendered as man the singer, man the musician. Do we find poetry so compelling because of its inherent lyrical, hence musical, quality? I wonder.

Should not the music of place and its other-than-human beings come first in our learning, rather than unsung words alone, or songs unconnected to place-beings? One thinks of children growing up on television and radio jingles, or nursery rhymes set to music but not to place-beings. Words are too perilous to be uttered out of a genuinely earthy context; they are too inherently powerful to be left unmoored, unaffiliated with place and the sentient beings there. Insofar as our words, both spoken and written, are not rooted in precise place and learned from such place—I would emphasize learned *in song* from place-beings—such free-floating, detached speech becomes dangerous and, often, destructive, even if inadvertently so. For such disoriented, strictly human-oriented speech has always tended to succumb to fear and, worse, its corollary, mendacity.

"Once in his life a man ought to concentrate his mind upon the remembered earth, I believe." So Momaday begins a wise and lyrical and oft-quoted admonition in *The Way to Rainy Mountain*. "He ought to give himself up to a particular landscape in his experience, to look at it from as many angles as he can, to wonder about it, to dwell upon it. He ought to imagine that he touches it with his hands at every season and listens to the sounds that are made upon it. He ought to imagine the creatures there and all the faintest motions of the wind. He ought to recollect the glare of noon and all the colors of the dawn and dusk."[14] Above all, I would say, he ought to learn its songs.

~~~~~~~~~~~~~

And the song of historical consciousness? It has no song. History uttered and history written have replaced the songs of where and who that are taught by a singing earth. Years ago, as an adolescent, I underwent the academic rite of passage of reading Homer's *Odyssey* and *Iliad*, very powerful and, in some senses, mythic works. One thinks of Odysseus's heroic journey: the hero's death and rebirth expressed in mythologies the world over. It is an anthropic version of the bear's story. A few years ago, visiting with the Canadian poet Robert Bringhurst, I learned something else about these two poems. One day at lunch, Robert began singing one of them—in Greek. He had been talking to my wife, Nina, and saying that it didn't matter if one could not understand the language (which Bringhurst does);

he said these epic poems were meant to be heard as song. Merely to read or speak them was to deprive them of their greatest power, he declared.

I thought then of Thucydides and Herodotus, the founders, as it were, of the discipline of history: a narrative ordered by sequential dates, a narrative derived from empirical evidence rather than hearsay or speculation, a narrative purged of the fabulous and the supernatural—the antithesis of the Homeric epics. Thucydides' *History of the Peloponnesian War* and Herodotus's *Histories* were declaimed but, as far as I know, not sung. An ominous difference, I think.

The Greek rationalists who scorned the singing poets substituted one kind of truth for another; they exalted the truth of written history (*logos*) over the very different order of truth we call myth (*mythos*), which was sung. Bringhurst explains the difference in a footnote about the fourth-century emperor Julian the Apostate, "who tried to reintroduce paganism . . . to Rome long after Christianity had taken hold. One of his advisors was a pagan neoplatonist theologian named Saloustios, who wrote at that time a little compendium of Greek mythology. 'These things never happened,' Saloustios says in his preface, 'but they always are.'" In the ancient Mediterranean world, myth-songs were understood to mean "*ageless truth.*"[15] By speaking or writing the stories of strictly human events, bracketed between two dates, Thucydides and Herodotus, Xenophanes and Plato, and the host of historians and philosophers who followed them abdicated the realm of the larger, more inclusive story—the timeless story of myth, that always is and in which all is self—for the walled-in, human, self-nonself story that once was.

And yet, even by Homer's time and in his rendering (or that of whoever composed the *Iliad* and *Odyssey*) Greek mythology had largely lost its original mythic base. I would maintain that those original, primal stories, the ones in which mythology itself must have taken root, were probably the songs that hunter-gatherers learned in dream, in vision, in meditation, in ecstatic communion with singing place-beings: the human harmonics of bear song, wind song, prairie grass song, osprey and owl and loon song, and so on. The songs humans perceived in spirit of other beings and that were also then the songs of place, instructing the attentive listener on where he was and, from that, who he was: the "image of the world . . . engraved on the architecture of the spirit."[16]

〰〰〰〰〰〰〰

I have oftentimes glided my canoe along the margins of beaver ponds in the north woods, where, like Thoreau, "self-appointed inspector of snow-storms and rain-storms," I survey both dam and lodge, or their remains, and attend the strategy used in felling birches and aspen and other tree-kinds, all the vocation of the Old Man's nation, *Castor canadensis*—our alienated name for these swimmer people.[17]

I have stood as still as I could by a marsh in a certain Maine woods watching a pair of beavers repair a dam with mud and sticks expertly inserted into the failing structure. Watched them glide through water, dive, resurface farther on, with great V stretching out behind. I nearly intercepted one while canoeing near dusk on an Ontario lake several years ago. It suddenly raised its rump and thwacked the glassy surface with gunshot report, and vanished.

By my bedside I confess I keep no books, but a stick of birch fetched from an active lodge on that very same lake—stick stripped of bark and small, succulent branches, all cleanly chewed off, leaving but toothy signature. Occasionally I sit alone with that stick, feeling it gently with thumb and forefinger. I bring it near and scrutinize *Castor*'s art, *Castor*'s imagination. My mind sometimes wanders into the ledgers and journals, birchbark canoes and rude log outposts of the Canadian fur trade; *Castor* and *Ondatra* (muskrat), the Old Man's nation and the Little Man's nation, were the topics of those ledgers and journals, St. Lawrence–bound cargo in those monstrous canoes. Who knows how many *Castor* were captured in leg-hold or spring-pole traps or drowned in these diabolical contraptions set beneath the water *Castor* himself had so carefully husbanded, all to furnish fine felt for European hats? The swimmer people continue to be made into coats and hats and other assorted bits of outerwear. What was the mind, the logic that imaged these swimmer people in such a manner as to imagine and pronounce their wholesale slaughter? What kind of dreaming conjured up the steel trap? How could Five Nations and Cree and Montagnais and Ojibwa and a host of other human persons who knew better have reneged on their compact of mutual courtesy with the Old Man's nation?

The questions do trouble a handful of scholars, but with few exceptions only for academic reasons, it seems. I, myself, have contributed to the debate—and watched and heard my words caricatured and emulsified in academic sophistry. Now, however, I have a different audience in mind, and the issue exercising me is more urgent, less academic, more apparent: our present-day speech and artifice of the animal beings that were once upon a time deemed worthy of respect by human persons.

I was raised in rural Canada, yet the only beaver I recall encountering and contemplating in my callous youth (for I was a *sport* hunter) was, quite literally, a cartoon. My father had a wealthy friend who owned a large construction firm in Montreal, called Beaver Construction; my childhood beavers were squat and quizzical toothy images adorning the doors of this man's massive diesel trucks. He dressed his beavers in overalls, if I am not mistaken, and I think had them clutching either spade or pickaxe. At any rate, they were working for him now, on his terms, just as he pretended to be working in the spirit of *Castor*, the company totem, in a sense. The company has flourished in the years since.

The image is cute, but a lie, monstrously mendacious in its depiction of the swimmer people. This is just one species of self-deception our culture employs in its traffic in animal beings: cartoon. There is another, equally freighted, illustrated best I think by a brochure I picked up several years back at a Maine tourist bureau, advertising the cunning services of a man retailing himself as a hunting guide up there somewhere. He documented his skills with a grisly collection of photographs showing black bears and Virginia deer eviscerated and hanging by the neck from a crude block and tackle gallows near what he calls the hunting lodge. Standing beside each twisting creature is its grinning executioner, displaying the very instrument (rifle, handgun, or bow) that did the job. The brochure uses the words *record* and *trophy* to give meaning and context to what has happened—and I am reminded of John Coetzee's hero of enumeration.

Take an automobile trip through any section of the United States and note the number of billboards showing a smiling, roly-poly pig wearing a chef's hat and urging you to eat its flesh (as ham, say) at some restaurant up the road. Cows in similar cartoonish manner invite hungry motorists to consume their delectables, as do chickens, lobsters, crabs, fish, turkeys—the list is long. Oafish bears wear slouch hats and moronic grins to recommend a "nature" or "fun" park up ahead; penguins sport a scarf and the same silly grin. The docility and foolishness of these representations of these beings is reinforced and lent credibility by the soul-crushing, soul-walling institutions we have built and which we call, euphemistically, zoological parks. One thinks of the wild boars that were enslaved and genetically re-engineered millennia ago to become the hogs of Iowa farmers, the Asian pheasants that have wound up as mass produced fryer chickens on Maryland's Eastern Shore, and the biogeographically elusive Asian ruminants that sired the cattle now stripping the cover off a thousand hills.

It is worth reminding ourselves that insofar as domesticated swine are "piggish" in their habits, they are so because we have cornered them into foul circumstances, both in habitat and artificially selected temperament. Similarly have we turned cows and chickens and sheep and other domesticated stock into dolts, where their human-engineered stupidity inspires our ridicule, contempt, and

humor. The interesting thing is how we harness these toys to perform as potent symbols and metaphors of "real" life, or so we imagine, incorporating their manner and handicap into our daily speech
to declare of someone that he eats or lives like a pig, of another
that she is bovine in personality, or that another follows others like
a sheep.

Let us recognize, forthrightly, that we have fabricated these
barnyard creatures—their demeanor, their living conditions, and
even to a large degree their very morphology and physiology. Each
in its own way is an experiment in genetic engineering. Put more
powerfully, they are all, collectively, a weird testament to our awesome and frightening powers of imagination. And yet they are not to
be confused with the genuine article, defined as propagating and
otherwise performing on its own economy, including its own volition. We would be closer to the mark to recognize these small monsters we blithely call livestock as but the perambulating guts, bone,
and flesh which neolithic man cajoled out of the immense plasticity
and potential inherent in ancestral genes. Not God nor nature nor
the universe, nor the animal beings themselves, but *Homo* the artificer did craft an item of fleshy technology and, antecedent to that,
an item of thought, for our own strangely conceived ends.

My objective is not so much to labor what countless animal
rightists have said already: that we have perpetrated something monstrous upon these creature beings. That pronouncement falls on un-
discerning ears. My purpose is more to reflect on the image we engage in, now, today, in casual speech and awareness, in the blizzard
of advertising daily swirling around us, and in the endless other oral
and visual renderings of animal beings, as twisted as their livestock
referents are. It is the full import of this image that engages me. I
include, as well, the image of *trophy* bear and *record* buck, where
bear is imagined as dangerous, criminal almost, and deer as man's
legitimate sport and protectorate; I include the meaning, too, of these
imperfectly understood creature beings whose vocation and errand

we vainly imagine that we, human persons, can properly adjudicate.

We cannot. For what we are doing in all of these cases, as disparate as they may seem—whether manipulating their DNA on the farm, cartooning them in speech and advertisement or other visual means, or making them sport or felons or wards in the hunt and public policy—what we are doing is lying about their authentic nature and meaning. In lying about these we become self-deceived about our own nature, purpose, and powers of judgment—this is what troubles me.

The really powerful force operating here is the way in which we have trained our collective and individual imaginations to feel, the way hands and lymphocytes will feel an unfamiliar object, these creature beings. This is the realm of cognition (image making), which responds, in part anyway, to another force: words. Speech is the other creative force operating here. Bring to mind the power of the word, and the world that was given coherence and lucidity, meaning, function, and texture by that Word. Humans partake of such immense and frightening force. Words, Momaday reminds us, are sacred. And anything sacred is fraught with danger, so all the mythologies of the world warn us repeatedly.

There is another lesson taught by mythology and familiar to hunter-gatherer societies, this being that words of place and place-beings, such as animal and plant beings, are instruments for triangulating our true selves within the cosmos. Here is something to think about carefully. We are in the habit of thinking of ourselves as zoological, botanical, cultural, and terrestrial geographers, who find, categorize, and map that which was lost, or not yet found. Continents, strange and rare plant and animal species, lost tribes. We glory in our exploratory and classificatory vocabulary and technology. But have we used the proper words to describe the things so witnessed?

Part of my job is to read early European accounts of encounters with far-off parts of the world, especially America in the sixteenth and seventeenth centuries. I pay special attention to the words, the images used to run the mind along the surface and then engulf familiar and unfamiliar structures into European cosmology. Here, sea cows, dugong, or manatee (it's impossible to be precise) become mermaids; the early French in Canada called beaver, fish (in part for liturgical reasons), and called Canada's resident peoples, *sauvages*: wild men, those without benefit of law, religion, government, or manners. Thus say the journals.

Words. It was with the gargantuan force of Christendom's words that the French uttered into being a vast mercantile empire in the northern reaches of this continent, this land thought by the Five Nations to rest on the carapace of a slowly paddling turtle. And when an irate Charles V notified his detested French rival that certain words of his Holiness, the pope, had actually bestowed Canada upon Spain, not France, Francis I replied dryly that he wished to see the will of the first man, Adam, giving instructions on how the world was to be parceled out.

Empires called forth by words. So Europeans sailed for America in the sixteenth and seventeenth centuries armed with words for the earth and its seas, its multitudinous animals and plants and birds, its sea life, its people. They expressed it all in words we now find quaint and groping, naive, often amusing, often wrong, and on many occasions bearing tragic or even ghastly results. Witness the rhetoric of Spanish conquest of the Caribbean and mainland in that bizarre document, the "Requirement," to be read aloud to the uncomprehending *Indios*, solemnly notifying them that the pope, God's vicar, had recently "made donation" of these lands to Spain (the same donation Francis had scoffed at)—words that instantaneously transformed the indigenes into tenants and subjects of the realm, much like Han Yu's words, which instantaneously transformed crocodiles into criminals.[18] Anything less than prompt compliance was treason, justifying the horror which the document then threatened to unleash. (*Conquistadores* became notorious for using hunting dogs to tear Indians apart.) The Spanish could "require" the indigenes to submit to their yoke for basically the same reason the Confucian governor could insist that crocodiles "agree" to quit the realm: both were speaking for a neolithically rooted principle of order, or time, to which "Indians" and "crocodiles" (both doubtless misnamed), together with the various and sundry other "creatures" of the universe, were to bow henceforth. Their submission was to the chosen people of a divinely calibrated history.

It is interesting that these explorers, *conquistadores*, merchants, missionaries, colonists, administrators, and soldiers thought they knew precisely where they were or where they were headed with their words: Columbus swore he had reached the Orient, the realm of Prester John (Balboa corrected him); Coronado marched a horde of men through the desert to the very edge of the plains in desperate pursuit of the Seven Cities of Gold (smoky, mud-brick pueblos, as it turned out); Cartier and his scurvy crew were beside themselves when Iroquoians casually introduced them to the medicinal holy grail, the Tree of Life, as he called it (merely northern white cedar, now adorning the front lawn of many a suburban home, bearing an abundance of ascorbic acid in its leaves). There is something vaguely comic about it all: oversized words both derived from, yet reinforcing, ill-fitting images of *Homo* vis-à-vis place. Whether comedy or tragedy, Europeans expressed America and its blessings into terms, a certain frame of reference, a being, even, that enabled them to dispossess summarily its aboriginals (many of whom were farmers, incidentally, though generally in combination with hunting and gathering) and to populate it instead with their own fauna and flora, pathogens and citizenry, not to mention their commerce and religion and other fanaticisms. And then they did the same thing in the South Pacific, in parts of Africa, and in fact a sizable portion of the rest of the globe.

As much as I would like to, I cannot change those events with my own words. I can, however, suggest a way to prevent ourselves from repeating the performance in future. And it has to do with the enormous power inherent in speech. In words. With some reflection, one begins to appreciate the terrible power that humanity's words have had in setting the course of what we like to call history. But what if those words are mendacious? Especially those words which purport to describe, to fix in one's own mind and that of others, things of place and place-beings?

It is precisely here that animal and plant beings are particularly relevant. Studies of surviving hunter societies make it clear, as already shown, that for the vast majority of its trajectory our kind has deferred to these other-than-human persons to determine its own metaphysical lines of longitude and latitude. There is the vision quest, and the song of place-beings; the refusal to split the mind and speech and artifice into self-nonself; and the richly pungent animalness and plantness of one's conversation and art/artifice.

For hunter societies, words of animal and plant, words of place, and the enabling powers of one's earth-derived artifice were transmitted by these beings themselves. It was essential to continue observing this manner of speech and artifice if one hoped to maintain a truthful and harmonious relationship to one's habitat. One of the great insights of hunter societies is that words and artifice of specific place and place-beings (animal and plant) constitute humanity's primary instruments of self-location, the computation of where, in the deepest sense, one is in the biosphere, using words and artifice that have accurately touched the place and these elder beings. For mankind is fundamentally an echo-locator, like our distant relatives the porpoise and the bat. "Her high sharp cries / Like shining needle-points of sound / Go out into the night and, echoing back, / Tell her what they have touched." Bat song, bat words. "She hears how far it is, how big it is, / Which way it's going: / She lives by hearing."[19] So, too, does *Homo*, and in a manner not too dissimilar from the bat. Only by learning, and scrupulously using, *true* words and *true* artifice about these things can one hope to become, in turn, a genuine person. Call it metaphysical echo-location, if you will.

Notice how jarring the word *true* is, here; it is a derivation of truth we are uncomfortable with—it seems specious. That should not surprise, for we today are heirs to a neolithic veracity which inverted the hunter *veritas*: in the neolithic, mankind would bring words and artifice to bear on the earth and other life forms as instruments for locating that which was construed as nonself, existing out there. Hunter *veritas* has always expressed the conviction that the thing seemingly out there was not out there at all, but part of the same fabric as ourselves, and hence, within—a piece of cosmic insight that has since been spectacularly confirmed by evolutionary biology, molecular biology, physics, and Freudian and Jungian psychology. Christian society, emblematic of that neolithic inversion, would militantly believe that its messiah, Jesus Christ, spoke the truth when he reportedly identified himself as the referent for truth—a chilling disengagement from the earth. Locate yourself in me, not place and place-beings, was the message of this distinctly neolithic savior. Centuries later, Europeans would wield that Christ-like vocabulary and technology of nonself place and nonself beings to implant their persons and institutions in parts of the globe where indigenes spoke and manufactured as though all was still self. But that, as they say, is history—water under the bridge.

Not quite. Water still flows under that bridge; we continue to speak of place and elder beings in such fashion. If *Homo* is in truth an echo-locator, and if, furthermore, it is true that place-beings (animal and plant) are essential referents in the process of self-finding, then surely we are lost. To be mendacious about other-than-human persons springs back upon us to make us mendacious about ourselves. When I cartoon bear and beaver and even pig, I disorient my true self; when I slaughter wildlife for trophies and records, or speak of myself as the earth's steward, I disfigure my true image, my true being. The underpinnings are wrong, the relationship turned inside out. Words are the enchanted mirror: when our words lie, the reflection deceives.

∿∿∿∿∿∿∿∿∿

The Amazon jungle: the popular mind imagines it as a paradigmatically fearsome place ("Nature, red in tooth and claw," wrote Tennyson). Yet my wife, Nina, lived in it, in a vast Peruvian national park the size of the state of Connecticut, for two years (all told) in a tent. She bathed in lakes with caymans and giant, five-foot-long river otters that hunted (fish) in packs, walked paths patrolled by jaguars and dangled over by boa constrictors—and was never once harmed or threatened by any of these myriad creatures. She lived there without benefit of rifle (guns were in fact forbidden by the Peruvian government), conducting research on tropical bird ecology. Though that was years ago, it remains her favorite place of the mind, and she would go back there at the drop of a hat. So much for the much-vaunted terrors of the jungle. It is humankind who terrorizes the jungle, not the jungle terrorizing us—except in our imagination. And yet how much of our self-congratulatory sense of civility and refinement and progress, even, is founded on this patently fictitious image? Even the word is wrong: it is tropical rainforest, not jungle.

Or take wolves. Wolves, since the dawn of pastoralism, have suffered from a vicious press. Let it be understood that wolves are not dangerous to humans, period; wildlife biologists have known this for years. It is humans who are manifestly dangerous to wolves, yet we commonly refuse to think of it that way. Wolves don't attack humans, Jack London's terror-stricken fiction notwithstanding. They hunt moose and caribou and musk ox and so forth—but that, in fact, is not my business. Not part of my economy. If I have presumed to make it part of my economy, then my economy is misconceived, misallocated. For that is wolf-moose economy solely, and my thoughts and efforts on the matter are transgressive. And when wolves prey on domestic stock, then it's time to rethink the very founding principles of the neolithic (not much chance of ranchers doing that), and, while mulling that over, add legitimate wolf predation into the company's books.

It is high time we took wolves off our nightmare list. The same goes for rattlesnakes, alligators, mountain lions, and a host of other carnivores. As for the handful of creatures that will, when hungry, seek to make a meal out of my body (which, contrary to popular opinion, is not a 24-hour, 365-day a year craving), it is merely incumbent upon me to stay out of their range. The solution is not to insist on occupying their territory and then attack them headlong with words and weapons. That is folly, at several levels. To malign such creature beings, again, inevitably results in my deceiving myself about who I truly am.

And yet this exercise in self-delusion has been the stock-in-trade of nascent civilizations the world over, and it continues to function at the very heart of our own today. Animal beings have always been fundamental beacons for human identity. When I see, in the iconography of the earliest farming civilizations of the Old World and the New, scenes depicting animal assaults upon humans, I know that the great lie, the great witchery, has been set in motion. It is there in all the cradles of civilization. The Christian tradition would pin the witchery on the serpent, and look for a savior. Truth is, it was the serpent being who was lied about, and it has ever since been crucified for neolithic man's sins.

~~~~~~~~~~~~~~~~~

Jesus Christ pronounced himself the truth, the way, and the life. Under an iron-clad neolithic paradigm, such messianism may not be an unreasonable proposition. My argument is not with Christ's presumption but, rather, the premise and context into which he came: the neolithic. That, I think, is where the fundamental disease of the world lies: in the neolithic turn of mind which, to our complete amazement and consternation, is methodically destroying much of this earthly habitat while showing no real hope of turning off the process. Witness the paradox of all of us in America sporting snappy bumper stickers and ample rhetoric declaring ourselves earnest ecologists, while each of us (myself included) continues to merchandise to death the very thing we talk about preserving.

If humanity is going to find a way out of this enchanted realm, we will have to first *speak* an aperture, then from that creative word follow a course vastly different from that preached by all the messiahs, religious and secular, who have pitched their message within an agricultural or pastoral metaphysic. My time (my past, present, and future) lies right here, smack on the back of this slowly paddling continental plate, not in some fantasy heaven or equally fantastic hell or any other Middle Eastern narrative. My being, my real and true economy, is in the humus of forests and parklands and open plains, in spiritual commerce with beaver and bear and deer and other animal beings resident there and performing so ably, so brilliantly, within those myriad communities of lands and seas and airs. There stands that shimmering biped, *Homo*—and there is *Homo*'s salvation.

The aperture will begin to form when we change our speech of place and elder beings, when we speak differently of these things. Speak, now, as hunters knew, truthfully. For we can only emerge from this forest of our own fashioning, only undo our self-inflicted, self-spoken state of being lost, by becoming, first, genuinely human. As jarring and outrageous as it may appear, we are not now true humans. If our speech lies, then we are something other than the genuine article. The thought makes one's flesh crawl—but think it anyway. The horror is that we have uttered ourselves into barnyard-billboard-trophy-conserved, and mostly frightened, monsters.

〰〰〰〰〰〰〰

Earth speaking earth,
singing water and air,
audible everywhere
there is no one to listen.[20]

# 5

## "We have come to where history ends"

∿∿∿∿∿

### Breaking History's Hammerlock on Our Imagination

Dad was a preacher who introduced himself loudly as a Minister of the Gospel and who fed his kids on two books. The Bible, naturally, from which we partook aloud each evening, a verse apiece, one or two chapters, round and round the family of seven. "Seven equals Divine Completeness," intoned father, who would then expound on the mysteries of the Word. I grew up attentive to the majestic cadences of the King James Bible. Being saved was to reproduce the frightful journey of Christian, Christ's Ulysses, pressing on to the Celestial City, fleeing the City of Destruction—Bunyan's *Pilgrim's Progress* was the other chief primer in those formative years. I cannot say whether my father meant me to or if it was just my youthful imagination and Christian zeal that caused me to associate Bunyan's landscape with the large tract of land where I was raised.

Wildwood, as we named it, was a wondrous place of woods and fields and marshes, bordering the lower end of the Ottawa River, where it splayed out as the Lake of Two Mountains. From here, the river joined the queen of North American rivers, the St. Lawrence. It has been my distinct privilege to have lived on both of these magnificent watercourses, to have known them in all their temperaments and songs and shapes. It is no accident that I have returned to each as an adult, in the habit of professional historian, to contemplate the ways in which each has borne North American history on its waters.

The Ottawa was the chief geographical focus of my early life; its seasons and moods seemed to set the terms of the landscape bordering it. Wildwood. I can still daydream it into existence, in that strange realm of near-visual imaging that is human memory: the trees I climbed and sometimes tumbled out of, built forts in, and swung from as in Robert Frost's poem; the sugar maples, tapped for their sweet fairy liquor, clouds of steam issuing from sturdy kettles on the woodburning stove, yielding the chestnut-colored syrup, that food of the Wild Wood—to eat it was to be possessed by its spirits; the trees the men sawed, split, and cord-stacked for firewood, to run that great furnace of a stove and generate the heat throughout those desperately cold yet utterly magical winters.

My dad kept a small staff of workmen on hand year-round, since in the summer months he ran a camp on a portion of the land and the men constituted the carpenters and plumbers and general handymen who built and maintained it all. For the most part they were middle-aged to older, down on their luck, usually veterans of one war or another, often washed up by alcohol; he located them through the offices of the Salvation Army. Some were certified pirates and not infrequently ex-cons; all were exotic in one way or another, and they taught me much. There was the former trapper who with such grace honed my paddling skills; the world-traveled tree surgeon with marvelous dragon ring, who coached me in the art of blowing the nose cleanly without benefit of handkerchief; and the fellow who chewed and swallowed razor blades—I swear it—or, if preferred, light bulbs, a performance guaranteed to awe any audience (which it genuinely did, though I refused to try it myself). All these and more essential lessons were patiently conveyed to an awkward youngster with big eyes and a gullible nature.

"The men," we simply called them, worked the land; though reeking of alcohol, they labored with sober industry to transform woods and field into a rarefied form of Christian piety and delight. As dad was wont to put it, ever mindful of his enemy, the Pope, in this bastion of Catholic idolatry (can anything approach Protestant invective for its resonance and richness of imagery, I wonder): "Make the place look *Protestant*." They did, I suppose. And that became the history of this place in the middle decades of the twentieth century: Harold George Martin making it look Protestant.

I was a fervent Christian as a boy, with a mystical turn of mind, and my relationship with the land and river was ambiguous. I had the sense there was something vaguely chaotic about the place, which dad was constantly battling. He wanted sweeping lawns and formal gardens; stately trees in a tidy row here, randomly spaced there; dormitories, a chapel, and a workshop, with no leaks; a harbor for the boats; a sandy beach. But the river and its landscape had other processes in mind, and the two, my father and the earth, seemed in perpetual conflict. So the priest-king, with his divine mandate to impose the godly will, wrestled with the things of the earth—and I grew up a bystander. I loved the Lord, but found I loved the things of the earth, too: the dog-tooth violets and trillium in springtime; the river in flood, filling in wooded hollows and valleys, drowning roads, while great carp mated throughout this new watery realm; ice heaving, booming, and cracking as the river changed shape, taking dad's cherished dock with it; howling blizzards leaving behind mountains of drifted snow; ice storms glazing branches and twigs with fairy glass. The place dazzled me, and I was often chastised for daydreaming within it. Dad knew I should be fighting the Good Fight.

He was making, he explained to those who asked, a place where sinners might come to know the Lord, and so assure themselves of the eternal glories of heaven. Nothing was spared in the effort to hack out a kind of heavenly geography, or sacredness, from this portion of the river and its woods: the celestial will stamped on the profane earth. Dad was a missionary in the spirit of the neolithic titans, an architect of history in its original and purest expression. Thus, the landscape itself became the object of Christian, historical conversion, a task assigned to the derelicts and other riffraff on the payroll, working under the vigorous direction of Doc (as they called him). French-speaking Roman Catholics were the human subjects of his endeavors, and for this enterprise money poured in from concerned believers in Canada and abroad, responding to "The Christian Home Hour" radio broadcast, endless "prayer letters," *Home Magazine*, and the occasional speaking tour. I grew up persuaded that our Catholic neighbors were idolatrous, and Catholic clergy positively satanic. Wildwood seemed under siege by the forces of Darkness; it didn't help that we were bordered on one side by the virulent French-Canadian nationalist Abbé Lionel Groulx, and on another by a community of Sulpician brothers.

Here, as a boy, I found my place and role in a time and errand I believed were guided by God's hand. Meanwhile the river coursed on its untamed ends, and the trees whispered their inhuman songs. Alone on walks I often puzzled over two robust footbridges made of piled stone, each choking off a short neck of marshland. Between them, the stone bridges connected our point of land to the property of the Sulpicians. The latter property included a large, heavily wooded island we always referred to as Indian Island; the maps showed it as Île-aux-Tourtes. We frequently camped on the island in the warmer months. It was noteworthy for its superb harbor, facing the Ottawa, enclosed by two large jetties of loose stone, made in identical fashion to our two stone bridges. Around the perimeter of the island were the remains of what had once been a wall, again of loose stone. The meaning and significance of this may have been known to my parents, but it remained a complete mystery to me until decades later, long after dad had sold the property to a developer and we had moved away.

As a historian I learned of the critical role of the Ottawa as the main artery into the continent's interior for the French- and then English-controlled fur companies in the seventeenth, eighteenth,

and early nineteenth centuries. Voyageurs had paddled and sere-
naded their thirty-foot freighter canoes, flotillas of them, up the
main channel of the lake that I had gazed and daydreamed upon as a
child. In those days the two hundred acres that would one day fall
into my father's hands formed a tiny part of a vast seigneury deeded
by the French crown to the Compagnie de Saint-Sulpice. About the
turn of the eighteenth century, the Sulpicians had erected a half-
dozen missions on or around the island of Montreal to minister to
converted "sauvages," including one, the short-lived Mission de l'Île-
aux-Tourtes, built at the tip of the Île-aux-Tourtes as it thrust out
into the Lake of Two Mountains, just before the river turned to
rapids and plunged into the St. Lawrence. Here lay the explanation
for the fine harbor, the stone wall girdling the island, and the two
stone bridges connecting all of this to the farther point of land that
would one day be known as Wildwood.

As all of this percolated into my consciousness, the ironies of
history began to reveal themselves. Centuries before, the Sulpicians
had also been missionaries, laboring over this identical landscape to
make it into a place receptive to their historical mandate, the divine
will, the will of godly history. They had endeavored to wall out the
river, wall out disorder, and create access where the land provided
none. The stone works were evidence of prodigious human effort,
probably done by the missionized Nipissing and Algonkin at priestly
incentive. My father called his religious organization Christian Homes
for Children; it occurred to me that his hated rivals, the priests, had
built their own version more or less on this same spot of land centuries before—their Christian Homes for Indians. The absurdity of
who was the legitimate Christian in all of this was palpable, as was
the absurdity that my father was ministering now, he believed, to
those who had ministered in the name of the same god to the majority population of the region centuries before.

Here was history being worked out in magnificent microcosm.
An exquisite illustration of the mind and program of history, it was

also a spectacular illustration of the lunacy of it all. The real events here were those of the river and its bordering forests, marshlands, and fields, the events I instinctively tuned into and refreshed my soul with as a child—the time of the river, the time of the earth on this sacred spot. There came a day when I chided myself for not knowing, in my youth, of the Sulpician enterprise and for being unaware of the role of the Ottawa in the fur trade. Only later did I come to believe that these events and monuments, such as they were, including my father's endeavors, were not relevant to the true nature of this place, its essential and embedded history, if you will. *That* story I knew as surely as anyone may know it, as a child: the sap running in its sugar maples as the first emblem of spring, the river in flood, the snow-blind cyclones of winter and clear gusts of spring, the silences and somberness of autumn, fox tracks in the snow. These were the events that set my historical clock, that molded my character, my vision, my sensations. And they are what I have continued to draw upon in the years since, wherever I have lived. This is the proper history of place and place-beings—and I was, and am, I now realize, one of those place-beings.

It is time, now, for me to come to terms with my rootedness in geography and its events, messages, and wheeling processes. Nothing I have done as an adult is worth, in my estimation, precedence over those earthly errands; there is nothing on my resumé, or my father's, or the Sulpician Order's, that is more important or timely than what that river and its natural borders are engaged in. Therein lie the narratives that empower, and they are the very best stories.

I would go back, except that the area has been so brutalized by a developer that it makes me weep. The river, now dammed by a series of walls upstream, is likewise out of character. I am witness to *true* history become mutilated. My stories and references have lost their attachment, their grounding, their confirming echoes. For, in my father's perverse reference of history, there came a time to sell it all, to another but different sort of developer—not surprisingly, someone who took bulldozers and chain saws to the place, who gouged out canals and built a score of luxury homes on the water, creating a pricey suburban refuge, from Montreal. The property is now connected to the island of Montreal by a huge bridge that is but one leg of the trans-Canada highway, which runs down the middle of the Île-aux-Tourtes, passing right over the Indian-made harbor and leaping into the air to span and make irrelevant the final end of the river itself.

In his old age, my father, living far removed from the place of his historical operations, is most anxious to tell his story of what transpired there at Wildwood. He wants to tell whoever will listen of papal and governmental persecution, of his tenacity and successes both in court and at winning souls. But there seems to be no audience for those stories. Guests listen politely and nod off. He has a book in mind, and at times I have sensed he would like me to write it; but it is the wrong story, and I am not engaged by it. The Sulpicians, on the other hand, have had their story committed to print, and God knows the fur companies have had theirs narrated many times over; and it is not inconceivable that someday, probably after he is long gone, my father's doings will turn up in a graduate student's dissertation. But they are, all of them, tales in folly; they are narratives conspicuously unhinged from the place where they transpired, tales of people possessed by ambitions and goals with hauntingly neolithic origins. My apocalyptic father, the mission-minded Sulpicians, the merchants running the great fur trade opera with the interior tribes—they were never interested in the expressions of the earth here, except as they might be harnessed for human-conceived ends.

It is the river and woodlands and marshlands that remain, though much transformed by the hand of man; the efforts of these previous architects of human-centered process are barely, if at all, perceptible. Wealthy Montrealers buy and build yet more homes on the Wildwood property. They enjoy seeing the river, and what remains of the woods. And yet it is important to understand that these things are not

mere scenery and that these people—all of us—are not drawn to
these things just to partake of them with our eyes and other senses.
These things of the earth are true history, in the sense of things that
endure and mold our nature, and we are drawn, I suspect, because
that sensibility has not been entirely bred out of us by historical
conniving.

Growing older, year by year, my perceptions of Wildwood
changed, the whole thing taking on the nature of a trajectory: it
acquired *history*. A specious history, I see now, that was merely an artifact of my family's performance and my own ageing. I would not
presume that this sacred spot claimed anything of its own that one
should call history in the conventional sense of the word: that process which transpires on the *outside*, as though there genuinely
existed such a reality as the theater of the nonself which might be
witnessed by external, detached observation and processed by im-
partial reason—the friendly alien perspective. The landscape and
river and seasons were never outside of me; Wildwood was the magician, the Flute-Player, the trickster, who captured me with its primordial music, reshaping, transforming, and verily defining me *into*
the stories that are, I now realize, its *songs*, where I am both singer
and part of the song, truly, a citizen of place. True history becomes a
captivity narrative. All other stories—being unpossessed by place,
including not having been, literally, taught by place—are on the one
hand a theft and on the other epistemological, ontological, and phe-
nomenological nonsense.

For me, who has known and even taught the narrative of human-
connived history, the river and its borderlands are all that remain, or
ever were, of genuine, lasting interest in this tale. Despite my parents'
efforts to divert my attention heavenward, I imbibed those voices and
powers of the earth, and I remain to a large degree their handiwork.
The river is my deepest, most powerful, history. It possessed me.

~~~~~~~~~~~~~

Speech, says the poet, is "the antler of the mind"—a magical image of hunter-gatherer speech, yet how flat when applied to our own.[1] Man the orator, man the artificer. What are we to speak, and what are we to fabricate? Such an elemental question sets up resonances touching the very core of our being.

Humankind speaks and fabricates within its perceived context, a fact as valid for hunting and gathering societies as for neolithic societies and our own, today. Context explains all. Hunters, as amply shown in the preceding chapters, conceived a vastly different emergence and context for themselves than did neolithic peoples. Hunters drew upon animal and plant origins (totemism) while beseeching those surrounding other-than-human persons for illumination and even dialogue in the fine art of living—living not especially as a human being (an incomprehensibly artificial abstraction) but as any being that lives. Speech and artifice were quarried within this perceived context, partaking of its peculiar, unwalled structure. Here words and artisanry sought above all to convey a universality and comprehensiveness of being, of power, of identity—something not insularly human, in other words.

Civilized societies, on the other hand, as we have seen, decamped this universal dimension for the desperately lonely yet extraordinarily potent realm of mankind apart, through the agency—mark this carefully—of deities. The neolithic gods, through their prophets and priest-kings, trumpeted a new story for humankind; instead of the hurly-burly of earthy relationships, of myths, the gods talked of mission and destiny, of profound tragedy and soaring heroics, of man called forth from intimacy with the bestial and vegetal—in the Judeo-Christian narrative, of man and his friendly god together on a majestic journey through time and space. In fine, by inventing gods, neolithic societies contrived to invent history. Divinities were essential as the propellant in breaking humans free of the intense gravitational pull of mythology, with all it bespoke—an ingenious device in metaphysical rocketry. A god-ordained, predictable and fixed universe (at least the one in which humans now existed), wherein mankind was sole keeper and intelligence, became the cornerstone of a new cosmic paradigm which, in the West, under the influence of the likes of Newton, Descartes, and Bacon, eventually rendered the whole thing as one gigantic, well-oiled machine. Doubts as to the validity of the model began to creep in with the growing realization, during the nineteenth century, of the earth's troubled geological past, and indeed present, coupled with the new and vexing evolutionary philosophy. Poised as we are on the point of exiting this present century and millennium, the doubts have become only further compounded.

Today the stark truth of the matter is that "the universe / is not for man, it is / itself." As for us, "the eternal form eludes us—the shape we conceive as ours. . . . We are one of many appearances of the thing called Life," assures Loren Eiseley, and "not its perfect image, for it has no image except Life, and life is multitudinous and emergent in the stream of time."[2] Through the lens of paleontology and evolutionary biology we now

> gaze backward into a contracting cone of life until words leave us and all we know is dissolved into the simple circuits of a reptilian brain. . . .
> We are rag dolls made out of many ages and skins, changelings who have slept in wood nests or hissed in the uncouth guise of waddling amphibians. We have played such roles for infinitely longer ages than we have been men. Our identity is a dream. We are process, not reality, for reality is an illusion of the daylight—the light of our particular day. In a fortnight, as aeons are measured, we may lie silent in a bed of stone, or, as has happened in the past, be figured in another guise.[3]

The neolithic image of man apart, administering a world whose terms are etched in stone, lies hammer-shattered as an illusion, an amnesiac's dream. Perhaps, allowed Eiseley, "perhaps the primitives were wiser in the ways of the trickster universe than ourselves." Eiseley struggled mightily to take the spotlight off humans and our amateur yet disastrous sorceries, and to put it on nature: "Nature is the receptacle which contains man and into which he finally sinks to rest," he would declare. "It implies all, absolutely all, that man knows or can know."[4] Thoreau, Eiseley's inspiration, said as much in the oft-quoted line, "Shall I not have intelligence with the earth? Am I not partly leaves and vegetable mould myself?" "Water and earth / lurch, wrestle and twist in their purposeless / war, of which we / are a consequence, not an answer," echoes Robert Bringhurst.[5]

The gods who marched man out of the enchanted realm of myth and bewitched him, instead, with history are now effectively inarticulate, as far as most intellectuals are concerned. And yet their gospel—the consciousness of history—remains as intoxicating as ever. One might say that the woolly secular ideologies of the state have assumed the mantle from religious dogma, and in this sense it could be claimed that the gods remain operational in the performance and welfare of the state. Then, too, the deities are hardly dead for the rank and file, time-serving citizenry who throng the churches. Where once the deity had been reified, now human institutions and ideologies hold forth as the real, though a spurious reality, to be sure.

For there is, we now know beyond reasonable doubt, only the reality of the universe and, most urgently for humankind, the reality of this soiled planet. "Earth is the ultimate substance. / What is, is made out of earth. We / who climb free of it, / milkthistles, mallards and men, / are made out of earth which is driven by water."[6] None of which, I might add, is particularly charmed by mankind's monuments and endeavors. "The earth makes wraiths of us, prefers to see / the objects we conceived smashed into shards. / . . . Millennia create beauty, man merely adds his thought / and vanishes into air. . . ."[7] It is the manifesto of a new ecological consciousness. Thistles, mallards, and men are truly made of the same mud: fats, carbohydrates, nucleic and L-amino acids. At the microscopic level, the myriad living shapes which to the unaided eye appear unique, run together, like a child's watercolor palate, into the common molecules constituting life—life now viewed by the most discerning as a seamless tapestry of many dimensions. A biosphere of communion, of transformation, of translation. There is no tenable place for man apart or fixed man in this perception of our context, a context that is truly and utterly real.

We know all this; we've known it for decades. Some have known it much longer than that. We know, too, that human enterprise within this utterly real and true biospheric context has run amuck. In his posthumously published *Star Thrower*, Eiseley asked the question that haunts me: "Is not the real business of the artist to seek for man's salvation, and by understanding his ingredients to make him less of an outlaw to himself, civilize him, in fact, back into that titanic otherness, that star's substance from which he had arisen?"[8] Is it not the real business of the social scientist and scientist, as well?

But we do know man's ingredients, now, and the ingredients of all the rest of it, too. We know enough, at least, of the handshake of it all to be convinced of the folly of civilization's course. And yet we know of no effective means of arresting civilized man's trajectory— his mad compulsion "to wreak his thought upon the body of the world."[9] Even so, it strikes me that we now stand on the threshold of a major paradigm shift. The basic knowledge of biospheric communion is well in hand, and the alarms have been sounded over and over for civilization's profligate and destructive ways, while, moreover, the intelligentsia, at least, are no longer in the grip of deities of manifest destiny nor do they truly believe any longer in the ideological cant of the state. Things seem in place for a massive structural change.

One thing, I believe, thwarts it: history's hammerlock on our imagination. Historical consciousness itself has become perhaps our greatest enemy to true progress, the greatest obstacle to imagining ourselves and recalibrating our affairs in line with the new environmental consciousness. The latter is a primal consciousness, actually, for it marks the essential mind and speech and artifice of hunter-gatherer societies vis-à-vis place and place-beings; and that fills me with hope: there is a precedent among *Homo sapiens*. The problem, flatly stated, lies not with *Homo sapiens* but with neolithic *Homo*: farming, urban-based, and ultimately industrial societies.

"Far from consisting in change," genuine progress, declared the philosopher George Santayana, "depends on retentiveness. . . . When experience is not retained, as among savages, infancy is perpetual." Which Santayana followed with his now famous warning, "Those who cannot remember the past are condemned to repeat it."[10] In the popular rendering, *history* has been substituted for *past*, and, so edited, Santayana's dictum has become enshrined as the ultimate justification for history. Ironically, it is of course history itself which forgets. "I am ashamed to see what a shallow village tale our so-called History is," confessed Ralph Waldo Emerson. "Broader and deeper we must write our annals . . . if we would trulier express our central and wide-related nature, instead of this old chronology of selfishness and pride to which we have too long lent our eyes."[11]

History forgets. It forgets the truth of mankind's embeddedness in this earthy planet; it forgets that there is no interesting or sane account of mankind apart from this larger narrative. Yet, somehow, in myth and the quest for empowerment, the savages Santayana impolitely referred to remembered it all—spoke about all of that. They kept their eye on it all as they charted their lives. Consider the trickster, the animal-human exemplar who bungles his way through life and death, butting up against this or that situation, careening along and getting into all sorts of jams and scrapes and mischief. And yet always surviving. The trickster is a wonderful metaphor of animal (human) evolution over the eons. Imagine this character as a strand of DNA making its way through time and space, through one shape after another, trying on this and that. Such an individual (genome) has no particular mission except to survive. He is not malicious or particularly benevolent. He survives; he exercises his curiosity; he lives with gusto. Notice, too, that he lives within the terms and powers of creation; he is not master over them; he is not superior to them; he gets by with his wits, his guile, his blandishments, and even his foolishness.

Alas, my fellow historians and I are not trained to write in these embedded, long-haul biological and geological terms. Nor would most people consider this sort of thing real history, anyway; it's ecology, or evolutionary biology, paleontology, and geology, they would protest. And the multitude, along with their elected or self-appointed leaders, pulpit preachers, and newspaper editors, go on looking for "lessons from the past" from narratives which but invoke the neolithic empire of history, Emerson's "shallow village tale," Robin Fox's "brief episode in a temperate interglacial."[12] Actually, it's not just that historians lack the training to relate the long and, at the same time, the embedded view—though in fairness there are some notable exceptions, and, thankfully, their number seems to be growing with each new crop of Ph.D.'s. What disturbs me more is that none of us, I'll wager, has the courage to downright refuse to tell that "village tale" any longer. By telling it we resuscitate its fundamental terms: its peculiar premises and strange powers that stab us to our very core. Somehow, in the alchemy of uttering it—never mind whether one is critical or supportive—one is inevitably sucked into its vortex, becoming, paradoxically, both victim and collaborator. For we have lent it our voice, and the creative and sustaining power of that immensely potent instrument. The situation is not unlike coming home at the end of a trying day at the office and recapitulating to one's spouse what went on, the good and the bad: the event lives yet again, and its terms of being, its reality, are confirmed and are once more conferred life—in the end, both the narrator's and the auditor's.

Turning all of this inside out, we might ask what it is that we, in our time, find so compelling about historical consciousness? What are we trying to remember, anyway? What do the majority of us in the industrialized, bureaucratized nations find so engaging about history—the big picture, that is?

It is, I believe, the story of progress. We are proud of human progress, ascent from a misperceived paleolithic baseline. A great upward thrust out of what is popularly called the Stone Age, as though there were something constitutionally defective about the paleolithic. That argument no longer has merit, as I hope my earlier chapters have demonstrated. But, one might interject, if the hunter-gatherer state had not been somehow diseased or deficient, then why did certain societies abandon its constitution with life and embark on another course of engagement with the world? That, too, has been addressed already, but I would reiterate that the answer may well lie, finally, in a lack of faith. Infidelity: the gnawing fear that the earth does not truly take care of us, of our kind; the lie (which is what it is) that the world is not truly congenial to *Homo*, sapient *Homo*, with its unrivaled capacity for nightmarish imagination coupled with a stupendous talent for nightmarishly clever speech and nightmarishly ingenious artifice.

History, for the majority of us in the Westernized world, is the chronicle of mankind's great achievements under the banners of civilization. The conquest of epidemic disease, for example, in the last century or so; the ability to feed ourselves amply; the blessings of sanitation, medicine, food, and even education that developed countries have brought to underprivileged nations throughout the world. These are the sorts of things history, in its most comprehensive scope and application, is most proud to tell, in the popular rendering at any rate. I hear this litany all the time from my students. This is considered the bright side of history, the positive side of civilization.

"Big history" has yet another message to convey, again in the popular forum, of wars (like the two global wars in this century), of genocide (see, for example, Hitler's holocaust), and of institutions like human slavery. There are important lessons in these events which we must never forget, so one is reminded ad nauseum. Professional historians fill university libraries with monograph upon monograph seeking to understand their nature, their rationale at the time, so that we, in our time, might truly understand these terrible occasions and institutions from our collective past and perhaps, we hope, avoid repeating such blunders in the future. "Those who cannot remember the past are condemned to repeat it," after all.

Whether good news or bad, the story must be told. There is a grimness about our obsession and compulsion with telling the story. History, the cliché goes, teaches. It seems to be more than that, however; history has become a kind of colossal monument to humankind. We seem to want to declaim it to the heavens, other worlds even. It is the majestic, sometimes terrible, but still marvelous story of Mankind. We crave an audience to tell all of that to. How crushing to realize that this earth, with its teeming life, and this vast chambered universe do not listen and don't care.

But, then, it's the wrong story, and why should the earth be moved by inaccurate speech? It's a story wrongly put in all of its dimensions. As my colleague Robin Fox likes to put it, what we imagine as the plenitude of history is but a "problematical and experimental blip at the end" of a human trajectory that goes back at least five million years (more like six to eight million, it now appears)— "as though the last few thousand years are peculiarly privileged and the rest can simply be written off as 'pre-history'! This is pride, arrogance, hubris of a high order, and we are paying for it. Before Darwin, before we knew, it was perhaps forgivable. Now it is not even funny. It is the root of our self-destruction."[13]

Fox is well known for arguing that the human brain is optimally adapted to deal with (upper) paleolithic (i.e., small-band) encounters and arrangements of living: the hunting-and-gathering mode, in other words. In departing this environment of evolutionary adaptation, as he calls it, neolithic societies swung away from what was the real center of human nature and launched the species on what would become ever more wild, pendulumlike oscillations of aberrant, experimental group behavior—the very exercise we grace with the word *history*. History, as such, taxes the capacity and tolerance of a mind constitutionally unsympathetic—neurologically not wired—for such a lunatic enterprise and contextualization. And so we rebel, says Fox, in ways psychologists, sociologists, and reformers call aberrant, when in fact our rebellious ways are none other than an often pitiful attempt at self-healing and reversion to a species sanity.[14]

A story wrongly put, indeed. How about the good parts, mankind's conquest of disease, his conquest of hunger, and the benefits of civilization carried to underdeveloped parts of the globe?

Medical historians have written volumes on the centrality of the public health movement in the last 150 years in genuinely and spectacularly curbing such mass killers as tuberculosis, cholera, and typhoid. They hold up the germ theory of Louis Pasteur and Robert Koch and others as the opening wedge in the astonishing development and application of antibiotics and vaccinations. But what they don't talk about is the fact that it was the neolithic itself which created the monster—this new and unprecedented age of epidemic disease—that ultimately enveloped the entire world and was brought under control only when sanitary measures, widespread vaccinations, and, to a lesser degree, antibiotics, were deployed. Actually, it's slightly more complicated than this: there was also the natural, evolutionary attenuation of many of these contagions, probably combined with elimination from the gene pool of those individuals who were genetically most susceptible (crudely, a kind of biological editing). Laymen and, I'll bet, most historians labor under the misconception that the great infectious killers of past centuries have always been with humans, when the truth is they are artifacts of mankind's domestication of animals (when you domesticate animals, you inadvertently domesticate their diseases) joined with the creation of large aggregates of people (towns and cities) that proved ideal for the survival and propagation of many of these pathogens. Hunter-gatherers, scrupulously keeping animals at arm's length while maintaining themselves in small, relatively isolated communities, were simply incapable of creating such an epidemiological holocaust. The vast majority of diseases we write about in human history should be properly confined to the neolithic and postneolithic (industrial) blip of human history.

The same might be said for the conquest of hunger. Agricultural societies, with their comparatively immense size and stubborn refusal to curb population growth, and by their dependency on a limited number of artificially grown crops that are extremely vulnerable to drought, insect damage, and fungus rot, have brought on their own famines. Add wars and other large-scale social upheavals and you have several other elements contributing to the same inescapable result: either outright starvation or malnutrition.

But history likes to point out how, through technological skill and scientific tinkering, modern industrial and postindustrial societies have eliminated this specter. And that is largely true for the industrial nations themselves. The irony—or is it the punishment?—of this apparent achievement has been that virtually all of us blessed with unlimited access to food overfeed ourselves. We pay for our success with obesity, and the constellation of grave health problems attached to it. (Or we exercise madly, trying to fight off this disease of affluence.) Then there is of course the matter of the foods we eat that physicians keep reminding us are depositing plaques in our blood vessels and otherwise occluding and stiffening them, causing coronaries, strokes, kidney disease, and other pathologies.

The point of this tale is that the hunger and malnutrition we see human ingenuity eradicating (unevenly) in history were, in fact, creations of a neolithic approach to the world; it is a tale we cannot legitimately extend back beyond the neolithic into the hunting-and-gathering realm. The conquest of hunger is the conquest of neolithic-engineered hunger, though few realize this, so mesmerized are we by this "shallow village tale" which we fancy embraces the entire, primal enterprise of humankind. Yet this story, like so many, stops dead with the neolithic; it is, in fact, a very recent and scarcely universal narrative. Putting it less charitably, it is a disturbing exercise in forgetting—the larger, longer story of humankind. "And we are paying for it," for our forgetfulness, as Fox said. Paying for it with our obesity-triggered diseases and a runaway population, among other things. Nutritionists have learned only in the past several decades a bit of wisdom hunter-gatherers practiced routinely (though, as I have shown, not invariably): that we function best, physiologically, at a point somewhat short of complete satiety, eating only about two-thirds of what a normal, non-obese individual has the urge to eat. It begins to sound as though our neolithic tale about furnishing ample board is really a tale about providing too much food for a good many of us.

History also forgets that the so-called underdeveloped countries were made that way by the magic wand of the civilization that now pretends to be their benefactor. The inroads of disease and colonialism (including commerce with the alien metropolis) and the resulting cultural and social unraveling brought these people to their knees, to the pitiful, resentful, and rebellious state in which we, in the West, have regarded these ungrateful wretches over the past five centuries. They were feeding themselves just fine, and managing their relatively few indigenous diseases, and all in all were demographically stable before the Columbuses, Cortéses, Cartiers, Drakes, Magellans, and James Cooks dropped anchor, with their swarms of pathogens, rats, and imperial mandates. Again, history, as popularly understood, forgets this. In fact, is scarcely even aware of it.

All of this is Big History, as it is widely understood and savored. And what are professional historians doing, meanwhile? Few today seem willing to take the multivolume approach of Arnold Toynbee or Will Durant (to take the most widely known examples): the panoramic view and interpretation of it all. That kind of virtuoso performance worked when everyone remained convinced that history rightly consisted of the rise of the West. With consciousness raised about other parts of the globe and extended to ignored segments of even our own society, scholars now regard those magisterial syntheses as hopelessly parochial. What irony!

In the last quarter century we have produced intimate, sometimes numbingly detailed analyses of what seems like every conceivable aspect of society and culture, not just in the West but the world over. Wherever documents exist, historians are there sifting through them, looking for development, for connections, and, too, meaning for those involved. We have shown remarkable resourcefulness in appealing to other sorts of evidence to complete the picture, or see it from a different angle: ethnography, archaeology, statistics of every conceivable variety, psychological theory, meteorology, etc. Thus, collateral disciplines have been mastered and liberally drawn upon, a marriage that has produced a blurring of traditional disciplinary boundaries. So academic history has moved into anthropology (while anthropology and a few others have been shifting their weight to history), into sociology, psychology, archaeology, statistics, and geography, and it threatens to move also into biology, chemistry, physics, and who knows what other exotic realms. The discipline of history has been on the move for the past quarter century; the past masters from fifty years ago would scarcely recognize their craft today. Then again, one might say this for virtually every area of academic inquiry, including medicine.

Never before has academic history cast such a wide net. While most of us have cheered, others have expressed unease, wondering if all this is true, legitimate history? If our discipline is on the move, let it not stop now; may it learn to attend to the music of a different drummer, "however measured or far away"—in fact, to the strains Thoreau himself so clearly heard, and Emerson, too, though perhaps with less clarity. The song of the longest time, "the beating of the earth's dark millennial heart." *That* is our realm, our context, our place. Our story, our history. Surely it is madness to imagine that we can any longer afford the luxury of imagining differently. "We have come," wrote Eiseley with chilling prescience, "to where history ends."[15] I prefer to think that we have come to where the proper narrative begins. Can historians practice that craft?

As I see it, the real message of the past several decades is that we stand in urgent need of a new center of historical consciousness, one, this time, unlike the Marxist school, or social history, or ethnohistory, or what have you—one that does not forget. History must bite the bullet and drastically redefine its context—its memory. And when it does, the more discerning and courageous among that new breed of historian will realize that there are some stories that actually should be forgotten. Mostly, the narrow stories of horror. For memory stories are great power, a trait shared with myth. It was not an idle amusement that hunters believed myths to be alive; such narratives have the creative, and destructive, power of living things—like the power of the living earth. Memory stories can thus become an erroneous, hence dangerous, guide to future behavior. Consider the Cold War and nuclear arms race stimulated by our endless wallowing in the experience of World War II. We repeat what we remember —but, I plead, we are remembering a deranged episode that occupied no more than a fraction of a second in the whole experience, and proper narrative, of humankind. By dwelling on these horrible events we condemn ourselves, and our children, to living in their terms—or to not living at all.

The weird thing is that those daily recollected terms are ultimately artificial and imaginary; they are violently superimposed on the true, primal terms we in fact live by, whether we like it or not and whether we choose to recognize it or not: the terms of this earth. Mankind now serves two masters, and they are in direct opposition to one another. One is the amnesiac of neolithic historical consciousness, who, like Melville's Confidence Man, appears in many masquerades and beguiles us with the liquor of detached rationality, fear, and unsustainable dreams. The other is the great, brooding presence of earthy process, whose songs and multitudinous creations no longer harmonize with *Homo*'s neolithically derailed speech and artifice. We will have to emancipate ourselves from the former; the latter, being the truer and certainly more ancient, we will have to learn to speak and fabricate with once again, not in a perfect duplication of paleolithic ways, to be sure, but by the adoption of the same essential constitution of earth-being and earth-memory.

Many will respond with that oft-heard reply, But we cannot go back! To which I respond, But we never left—never left our true, real context. *Homo* is still here on this planet earth, abiding in our most fundamental and necessary nature by its fundamental and necessary terms. We left that context only in our fevered imagination. It all began as an act of imagination, an illusory image—most fundamentally an image of fear—and so the corrective process must likewise begin with an image. Let us relearn what hunter-gatherers knew to the core of their being, that this place and its processes (even in our death) always takes care of us—that *Homo*'s citizenship and errand rest not with any creed or state but with "that star's substance from which he had arisen."[16]

that some of us who happen to be Indians as well as scholars place the word *space* before *time* when we write. . . . It is no accident at all, because Indian traditions exist in, and are primarily to be understood in relation to, space; they belong to the place where the people exist or originated. . . . Indeed, some realities, most notably the sacred, have little meaning except in the context of their spatial referents. When shorn of these . . . they are likewise shorn of their moral force and a large portion of their range of meanings; hence, of their explanatory value. ("Some Concerns Central to the Writing of 'Indian' History," *The Indian Historian* 10 [Winter 1977]: 18)

Again, I suggest the above readings as background, a way of forming what I believe to be an appropriate context. From here one might want to move into some thought-provoking interpretations by scholars of various disciplines. Frederick Turner's "essay in spiritual history" (p. 7), *Beyond Geography: The Western Spirit against the Wilderness* (1980), has had an enormous influence on me. For Turner, "the real story of the coming of European civilization to the wilderness of the world . . . is the story of a civilization that had substituted history for myth as a way of understanding life" (p. xi). *Beyond Geography* is a tour de force in intellectual history. Read it with Morris Berman's *Coming to Our Senses: Body and Spirit in the Hidden History of the West* (1989), which covers much the same territory while decrying the rationalist bias in historiography. See chapter 3, "The Body of History," in particular, wherein he urges historians to "take a leap to what we might call *corporéalité*, a visceral approach to history that puts the mind and body back together again" (p. 134). It is another innovative and courageous book. For the first rumblings of this battle between the written *logos* of the philosophers and historians, and the *mythos* of the singing poets, see Marcel Detienne, *The Creation of Mythology* (1986). Herbert Schneidau describes the Hebrew hostility to mythology in favor of a sacred history in his *Sacred Dis-*

*content: The Bible and Western Tradition* (1976).

Frederick Turner talks at length about the spiritual bankruptcy that overtook the early, primitive Church, yielding an institutional leviathan that by the time of the Crusades was appealing to violence to jump-start its failed spiritual engines and restore lost credibility. The habits of violence and fear-mongering were soon shipped to the colonies and unleashed there, among indigenes and on the landscape itself. And so, laments Turner, was a perfect opportunity for genuine spiritual renewal surrendered to a neurotic historical vision. Instead of becoming possessed by America, say, or Polynesia, or any of the other neo-Europes (Alfred Crosby's term, in *Ecological Imperialism*, 1986), Europeans, with notable individual exceptions, insisted on possession. Richard Slotkin, in *Regeneration through Violence: The Mythology of the American Frontier, 1600–1860* (1973), gives a magisterial survey of this phenomenon (summarized in his "Dreams and Genocide: The American Myth of Regeneration through Violence," *Journal of Popular Culture* 5 [Summer 1971]: 38–59).

Somewhere in here we slip into the fundamental nature of knowing (epistemology) and reality (ontology), for it is clear we are dealing with two massively different structures of thought, even of being: the mythic and the historical (or rational, or empirical), one might call them. Both systems are vast, exceedingly complex, and difficult of analysis. Michel Foucault is especially worth reading for his brilliant insights into what we think we know by what is called history and anthropology. See his essay "The Anthropological Sleep," in *The Order of Things: An Archaeology of the Human Sciences* (1970), for a biting critique of anthropology as an obstacle standing in the way of a new conception, a new philosophy, of mankind. The anthropological point of view must be dismantled, Foucault maintains, before we can begin to imagine ourselves differently and *think* again. See, as well, his *Archaeology of Knowledge* (1972). Anthropology's epistemological shortcomings are further laid bare in Adam Kuper's *The Invention of Primitive Society: Transformations of an Illusion* (1988) ("the theory of primitive society is about something which does not and never has existed," p. 8), and Johannes Fabian's *Time and the Other: How Anthropology Makes Its Object* (1983) ("it is by diagnosing anthropology's temporal discourse that one rediscovers . . . that there is no knowledge of the Other which is not also a temporal, historical, a political act," p. 1). Lucien Lévy-Bruhl's *Primitive Mentality* (1923) was seminal in this kind of thinking, sounding the alarm that aboriginals think profoundly differently from the way we do—not childishly, he cautioned, nor foolishly, but out of a different episteme. One can gain considerable insight into this realm of thought (called by some prelogical) from Claude Lévi-Strauss's *Savage Mind* (1966) and *Totemism* (1963). Chapter 9, "History and Dialectic," from *The Savage Mind*, is a potent essay on the limits of historical understanding. Hayden White's *Metahistory: The Historical Imagination in Nineteenth-Century Europe* (1973) cuts to the heart of the matter when he argues that the historical narrative is invariably circumscribed by the conceptual limitations of language. This "precritical" tyranny of language renders all historical narratives essentially poetic exercises, so that in the final analysis, according to White, there can be no truly objective history; one in fact chooses one's school of historical interpretation on purely moral or aesthetic grounds.

A sense of past, present, and future—time as a segmented linearity—is organic to our English language, indeed to Indo-European languages in general, observed Benjamin Whorf, who noted its absence among the Hopi, a Pueblo society in the American Southwest. Whorf's study of linguistic differences led him, and others since him, to suggest that such fundamental divergences in ways of speaking of the world are an enunciation of mutually incompatible, even mutually incomprehensible, ways of conceiving of the universe. See Whorf's *Language, Thought, and Reality: Selected Writings*, edited by Carroll (1956), and Dell Hymes's introduction to *The Origin and Diversification of Language*, edited by Swadesh and Sherzer (1971). Without fluency in the language of the Other—the classic object of anthropology, the reformist goal of history—one can scarcely hope to penetrate that intellectual sphere. Too often our scholarly habit has been to speak on behalf of these people, forcing them into our structure of thought while remaining oblivious to theirs. Let two illustrations of this scandalous practice suffice: Eric Wolf, *Europe and the People without History* (1982), and Francis Jennings, *The Ambiguous Iroquois Empire* (1984). See my critique of the former in "An Introduction Aboard the *Fidèle*," in *The American Indian and the Problem of History* (1987), and of the latter in *Reviews in American History* 13 (March 1985): 14–20.

For a model of cross-cultural understanding, as well as a dazzling analysis of the collision of history and myth in the British usurpation of Polynesia, I highly recommend Marshall Sahlins's *Islands of History* (1985). Sahlins's main purpose therein is to show how the structure of the Maori, Fijian, and Hawaiian thought world was recast as a result of this foreign intercourse; yet along the way he has much of value to say on the nature of history, anthropology, and myth. See, as well, his earlier *Culture and Practical Reason* (1976), "an anthropological critique of the idea that human cultures are formulated out of practical activity and, behind that, utilitarian interest" (p. vii). "All *praxis* is theoretical," pronounces Sahlins in *Islands of History*. "It begins always in concepts of the actors and of the objects of their existence, the cultural segmentations and values of an *a priori* system," which is itself "cosmological" (pp. 154–55). And we are back to the nature of knowing, as in knowing the earth and ourselves as human beings when we insist on being historically minded.

Returning to the imaginary Other, collateral reading in this evolving category—a double-edged sword, as Hayden White makes clear—is found in *The Wild Man Within: An Image in Western Thought from the Renaissance to Romanticism*, edited by Dudley and Novak (1972) (see especially "The Forms of Wildness," by White), and White, "The Noble Savage Theme as Fetish," in *First Images of America: The Impact of the New World on the Old*, edited by Chiappelli (vol.

1, 1976). Arthur Lovejoy's classic, *The Great Chain of Being: A Study of the History of an Idea* (1942), explains the larger scheme within which the ideas of *savage* and *primitive* came to fruition. These are applied systematically to the American Indians in Roy Harvey Pearce's *Savagism and Civilization: A Study of the Indian and the American Mind* (1953, 1965). As Turner, Schneidau, Berman, and others take pains to show, this assault on other, newly contacted societies must be understood from the perspective of a larger and concomitant denaturing of nature itself. For further insights into this, one might consult Clarence Glacken, *Traces on the Rhodian Shore: Nature and Culture in Western Thought from Ancient Times to the End of the Eighteenth Century* (1967), and Keith Thomas, *Man and the Natural World: A History of the Modern Sensibility* (1983) (chiefly a summary of nature appreciation in early modern England).

Perhaps what is needed is a cosmic lodestone that is miraculously—is this possible?—free of historicism, anthropologism, rationalism, relativism, positivism, structuralism, or any other cant, ideology, or agenda, if we are ever going to stop going round in circles wondering who this creature we call *Homo sapiens* is. Paul Shepard, philosopher and population biologist, comes as near as anyone to furnishing such a universal reference point. Reflecting on history's limitations, Shepard writes:

> Oddly enough, history is supposed to be that instrument for keeping the significance of the past, viewing the whole page or genealogy or growth pattern without losing sight of the present. Clearly there is something lacking in a tradition of history that fails to unify, to bring the past alive into the present. It is not only a matter of historians and their work, but of a cultural mode of perceiving that lacks a unifying sensibility for uniting these three octaves of the fourth dimension—the life cycle, group record, and deep evolution. (*Tender Carnivore*, p. xiv)

In a lifetime spent pondering the species, and not unfondly, Shepard has produced a series of books bearing such evocative titles as *The*

*Tender Carnivore and the Sacred Game* (1973), *Thinking Animals: Animals and the Development of Human Intelligence* (1978), *Nature and Madness* (1982), *The Sacred Paw: The Bear in Nature, Myth, and Literature* (with Barry Sanders, 1985), and a half dozen more.

Shepard has tried to convince us (especially in *The Tender Carnivore*) that our intellectual and motor skills are still those of our *cynegetic* (hunting-and-gathering) ancestors, as are our characteristic behaviors throughout the various stages of life (principally childhood and adolescence). The mind we use today was forged tens of thousands of years ago when we were omnivores attentive to the animals about us. "My thesis," says Shepard in the opening pages of *Thinking Animals*, "is that the mind and its organ, the brain, are in reality that part of us most dependent on the survival of animals. We are connected to animals not merely in the convenience of figures of speech . . . but by sinews that link speech to rationality, insight, intuition, and consciousness. It is not the same as thinking *about* animals. The connection is in the act and nature of thought, the working of mind" (p. 2).

*Nature and Madness* takes up Erich Fromm's question, "Can a society be sick?—The Pathology of Normalcy [in contemporary Western civlization]" (chapter 2 in *The Sane Society*, 1955), and, like Fromm, answers resoundingly that Western society is indeed deranged—that, as individuals, we moderns are mired in infantile postures of mind—because of our separation from nature. The neolithic, the searing monotheism of Judaism and the substitution of historical destiny for myth, the tyranny of Calvinist theology— Shepard singles out these and kindred alienations from the earth that occurred over the course of time and that serve now to doom each of us to a perpetual childishness of outlook on the world and our fellow humans.

*The Sacred Paw*, lastly, focuses on what undoubtedly has been the single most important animal for human cognition throughout *Homo*'s history: the bear. Reading this book, one begins to detect a corridor that can lead us into an earth sanity once more, not by means of a bear god or other similar fantasy, but the bona fide creature in the habit of the true wild. Shepard consistently roots humanity in the deepest time: the affairs of the earth, where human saneness (far more significant an issue than some screwball obsession with *survival*) begins and ends. And surely that—saneness vis-à-vis the authentic affairs of this planet—is a universal focus for all humankind. Read Shepard's books alongside *The Arrogance of Humanism* (1978), by David Ehrenfeld, another ecologist, and *The Comedy of Survival: In Search of an Environmental Ethic* (1972), by Joseph Meeker, a literary scholar. Humanism is our modern religion, a cult, and cosmic madness, argues Ehrenfeld. We must get beyond it, he implores—to the sort of centering and embedding Shepard describes, I would say.

Meeker, on the other hand, wrestles with the tradition of the tragic hero in Greek and Elizabethan literature (witness Achilles, Oedipus, and Orestes, and Hamlet), arguing forcefully that "the tragic tradition in literature and the disastrous misuse of the world's resources both rest upon some of the same philosophical ideas" (p. 63). These being (a) "the belief that the universe cares about the lives of human beings" (p. 47), (b) the conviction that humanity is superior to other life and non-life forms, and that man is "destined to exercise mastery over all natural processes, including those of his own body" (p. 48), and (c) the principle "that some truth exists in the universe which is more valuable than life itself" (p. 48). All are absurd and have been long since discredited, declares Meeker: "The world has never cared about man, nature has never shown itself to be inferior to humanity, and truth has never been revealed in its awesome majesty except perhaps in the creations of tragic literature. Tragedy does not imitate the conditions of life, but creates artificial conditions which people mimic in their attempts to attain the flattering illusions of dignity and honor" (p. 48). A much saner approach (I hesitate to use the word *ethic*) is illustrated by the comic mode in literature: the rogue, the picaresque. The trickster, indeed. If the earth has a plot, paraphrasing Meeker, it is surely comic: "organisms must adapt themselves to their circumstances in every possible way, must studiously avoid all-or-nothing choices, must prefer any alternative to death, must accept and encourage maximum diversity, must accommodate themselves to the accidental limitations of birth and environment, and must always prefer love to war—though if warfare is inevitable, it should be prosecuted so as to humble the enemy without destroying him" (pp. 46–47). When the old whoremaster in Joseph Heller's *Catch-22* is accused of being a shameless opportunist throughout the carnage and folly of twentieth-century European history, he simply responds, "I am a hundred and seven years old" (Meeker, p. 46). And the earth is older than that.

Humanism, one might argue, is the double helix of totemism unsprung into a Great Chain of Being; it works by purging from the human genealogy those animals and even plants that hunting-and-gathering peoples, and even horticultural societies, recognized as somehow rightly a part of being a fully articulated human person. Humanism, then, marks if not the antithesis then at least a perversion of what Shepard might call the cynegetic structure of thought. Let us examine that intellect, those arrangements of thinking and, by extension, living.

A good place to begin is with Marshall Sahlins's *Stone Age Economics* (1972), a truly revolutionary view of the hunting proposition. Unlike Shepard's books, where hunting peoples are shown as the template for modern man and woman, this one considers hunting economies strictly as successful or unsuccessful strategies for survival. Supported by ethnographic evidence primarily, Sahlins answers with an unequivocal affirmative. But to recognize this, he cautions, one must first strip away the accretions of biblical, neolithic, Enlightenment, anthropological, and current bourgeois prejudice which have for so long hidden this fact from view.

Sahlins finds hunters blessed with abundant leisure, few wants,

and a technology easily adequate to meet those wants, and well fed, healthy, and full of confidence in nature's bounty—so long as they kept a few key principles in mind.

> Hunting and gathering has all the strengths of its weaknesses. Periodic movement and restraint in wealth and population are at once imperatives of the economic practice and creative adaptations, the kinds of necessities of which virtues are made. Precisely in such a framework, affluence becomes possible. Mobility and moderation put hunters' ends within range of their technical means. An undeveloped mode of production is thus rendered highly effective. (p. 34)

One can find ample support for these startling conclusions in works such as Richard B. Lee, "What Hunters Do for a Living, or, How to Make Out on Scarce Resources," in *Man the Hunter*, edited by Lee and de Vore (1968); Don E. Dumond, "The Limitations of Human Population: A Natural History," *Science* 187 (February 28, 1975): 713–21; Frederick L. Dunn, "Epidemiological Factors: Health and Disease in Hunter-Gatherers, " in *Man the Hunter*; Francis L. Black, "Infectious Diseases in Primitive Societies," *Science* 187 (February 14, 1975): 515–18; Brian Hayden, "Population Control among Hunter/Gatherers," *World Archaeology* 4 (October 1972): 205–21; Robert F. Heizer, "Primitive Man as an Ecologic Factor," *Kroeber Anthropological Society Papers*, no. 13 (Fall 1955): 1–31; Robert M. Netting, "The Ecological Approach in Cultural Study," Addison-Wesley Modular Publications, no. 6 (1971); Elman R. Service, *The Hunters* (1966); Adrian Tanner, *Bringing Home Animals: Religious Ideology and Mode of Production of the Mistassini Cree Hunters* (1979); A. Theodore Steegman, Jr., editor, *Boreal Forest Adaptations: The Northern Algonkians* (1983); and in my own analysis of early European contact in eastern Canada, *Keepers of the Game: Indian-Animal Relationships and the Fur Trade* (1978). And this represents but a fraction of the literature on the New Hunters. Since the publication of *Stone Age Economics* there have been refinements and clarifications of some of the claims of affluence (the realization, for one, that commerce with Europeans, foreign pathogens, Christian missionization, and white usurpation of the land subverted and at times obliterated much of this mosaic), but the central thesis remains intact. As with all controversial subjects, however, it depends on who you talk to or read (see Carmel Schrire, "Wild Surmises on Savage Thoughts," in *Past and Present in Hunter Gatherer Studies*, edited by Schrire, 1984, for some leavening thoughts on hunter-gatherer research).

Such Stone Age economics were but the visible, sometimes quantifiable, expressions of a comprehensive metaphysic, or world view, that we can decipher to a certain degree through the study of mythology. Here, the name most commonly mentioned these days is Joseph Campbell's, the result, no doubt, of a recent and (as it turned out) widely viewed series of television interviews with Campbell shortly before his death. One might want to look at his *Hero with a Thousand Faces* (1949, 1968), a comparative study from around the world of the "Hero Monomyth," the so-called Heroic Journey: from life to death and back again, bearing the power (knowledge) of the gods to the community. Or there is *The Mythic Image* (1974), concentrating on the traditional centers of emergent civilization, again, globally, in the first several millennia B.C. and winding up with an analysis of the religious iconography of Buddhism, Christianity, and Islam. See especially chapter 2, "The Idea of a Cosmic Order," and the connection to calendars and the building of a sacred center (as in a holy city, with its temples and other religious monuments). What Campbell is actually talking about in these two elegantly written studies is the genesis and elaboration of *religious* thought as it built on a mythic base and gradually moved away from that base—as in moving away from totemism, for instance, and animism, to the realm of sky gods, priest-kings, and a sense of divine election, destiny, and mission.

In my opinion, a much better source for exploring the deep recesses of the mythic mind in its more primal manifestations is Mircea

Eliade. He is less readable than Campbell but more profound. In two remarkable books in particular, *Cosmos and History: The Myth of the Eternal Return* (1959) and *Myth and Reality* (1963), Eliade explains that myth is the story of the origin of things: the cosmos in its multitudinous expressions, including humans and their social and cultural institutions. Archaic people, as he calls them, recounted these narratives of the "strong" time to acquire those very same original powers in dealing with the mundane affairs of living and, not least, to restore this entropically inclined world to its original level of energy—the source, perhaps, of the anxiety over cosmic disorder which seems to lie at the core of all the agrarian religions. The power of revitalization lay in speaking the stories of creation, of the origins, and performing the rituals contained therein. The aim was to live as best one could in the powers, and proper relationships, of the original time. Deviation from this, says Eliade, meant losing contact with the cosmogonic moment and theater of events, which was to fall into duration and history, a hazard archaic societies carefully guarded against. All of this is contrasted very nicely with the emergence of historical fantasies among the Greeks, Jews, and Christians, fantasies of linear time that nonetheless kept alive elements of mythology, though often disfigured as the mandate of history.

I have appealed to pre-Columbian Mesoamerica for another illustration of the birthing and unfolding of historical process. Dennis Tedlock's award-winning translation of the Quiché Mayan *Popol Vuh* (1985), the Book of the Dawn of Life, as their Guatemalan descendants still call it, is a good place to begin one's reading. Here we see the ancient divinities and lordly ancestors being invented with animal and plant associations, but the whole cast is unmistakably godly-human. The true animals are either turned to stone or have gone into hiding or are pronounced treacherous, while the plants, most notably corn, are placed under the patronage of anthropomorphised deities. Withal, we witness the creation of sacred time in calendars charting the celestial machine that serves as the template for terrestrial affairs. Among the Maya, numbers are themselves divinities (see Barbara Tedlock, *Time and the Highland Maya*, 1982, on the divinatory powers of the *tzolkin* calendar). "For the Mayans," explains Dennis Tedlock, "the presence of a divine dimension in narratives of human affairs is not an imperfection but a necessity, and it is balanced by a necessary human dimension in narratives of divine affairs. At one end of the Popol Vuh [the beginning] the gods are preoccupied with the difficult task of making humans, and at the other [the time of Christendom] humans are preoccupied with the equally difficult task of finding the traces of divine movements in their own deeds" (p. 64).

Octavio Paz has written a virtuoso essay on the ghastly yet poetic pageant of Mayan historical destiny in "Food of the Gods," *New York Review of Books* 34 (February 26, 1987): 3–7. On the Aztecs, Children of the Sun and inheritors of this monstrous cosmology through the agency of the Toltecs, see Miguel León-Portilla's, *Aztec Thought and Culture: A Study of the Ancient Nahuatl Mind* (1963). Jacques Soustelle's *The Olmecs: The Oldest Civilization in Mexico* (1985) probes the earliest manifestations of this fall into history in this part of the world (though one might chart it, as well, in contemporary Chavín culture in Peru).

I have contrasted this model of creation with the trickster's world, suggesting, after Meeker, a more comic appreciation of the orchestrations of the earth. Stories of the trickster abound in mythologies around the world. Here he is Raven (Northwest Coast), there Coyote (Southwest), or the humpbacked Flute-Player (Anasazi), Nanabozo: the Great Hare (Ojibwa), or Glooskabe (Micmac). His name is legion, and his antics are known from pole to pole. He turns up in fairy tales and the myths of agrarian societies in yet a different array of masquerades and titles, but all are essentially the one archetype. Given the universalness of this individual and the ease of encountering his meddlesome ways in ethnographic and folkloric literature (even in one's own life, I sometimes fear), I have not thought it necessary to recommend any particular sources on the matter.

Before leaving mythology, there are two further books I should single out. One is an extremely useful collection of essays, *The Anthropology of Power: Ethnographic Studies from Asia, Oceania, and the New World*, edited by Fogelson and Adams (1977). The epistemology and phenomenology of *power*, what might be called spiritual power—the powers that all human societies have discerned emanating mysteriously from the earth and have sought to comprehend and channel—such is foundational to mythic thought. Within this esoteric realm of power is another embryonic idea: the power inherent in the *gift*. For insight into this principle in human cognition, I strongly suggest Lewis Hyde's *The Gift: Imagination and the Erotic Life of Property* (1979).

As mythology became colonized by religion, the concept of the gift was likewise subtly redefined. (Again, probably no issue has rivaled this one on the Richter scale of cross-cultural misunderstanding and resentment over the past five centuries, as indigenous, mythically based societies collided with European notions of property. See, for instance, Wilbur R. Jacobs, *Wilderness Politics and Indian Gifts*, 1950, on the confusion and, on the other hand, necessity of lavishing gifts on the natives in the British and French colonial empires.) Nowhere is this redefinition of the gift more evident than in the metaphysics of incipient agriculture, where the reference points of plant and animal usufruct are moved from the earthly to the heavenly domain, transferred from the thing itself to its divine keeper, inventor, and (above all) impersonator. Read the Old Testament, the Popol Vuh, or any of a variety of agrarian creation stories to witness the smooth rhetoric of this cosmic swindle, passed off as both charming story and, more ominously, organic law of creation. For the critical element of time in this (bearing in mind that the planetary and stellar keepers of time were deities of both earth and heaven), see Lynn Ceci, "Watchers of the Pleiades: Ethnoastronomy among Native Cultivators in Northeastern North America," *Ethnohistory* 25 (Fall 1978): 301–17; and D. Tedlock, translator, *Popol Vuh*. When I got my first good, telescopic look at that portentous constellation last autumn, I wept. Ceci's article won the first Robert F. Heizer Prize of the American Society for Ethnohistory by unanimous decision of the committee.

Curiously, most of the scholarly attention to the dawn of the neolithic is silent about the metaphysical plane and the connection with time. Anthropologists, archaeologists, and ecologists are fond, rather, of debating at length the demographics, (human) economies, and social ramifications of intensive farming, see-sawing back and forth over whether a ballooning population was what sparked a commitment to domesticated food sources or if a casual cultivation of foods quietly created a population boom which accelerated and reinforced what soon became full-blown farming. For the latter case, the question of first cause still remains to be answered. I recommend the following to anyone interested in the full flavor of the debate: Lewis R. Binford, "Post-Pleistocene Adaptations," in *New Perspectives in Archeology*, edited by Binford and Binford (1968); Bennet Bronson, "The Earliest Farming: Demography as Cause and Consequence," in *Population, Ecology, and Social Evolution*, edited by Polgar (1975); Mark N. Cohen, *The Food Crisis in Prehistory: Overpopulation and the Origins of Agriculture* (1977); J. V. S. Megaw, editor, *Hunters, Gatherers and the First Farmers beyond Europe: An Archaeological Survey* (1977); David Rindos, "Symbiosis, Instability, and the Origins and Spread of Agriculture: A New Model," *Current Anthropology* 21 (December 1980): 751–72; and Donald O. Henry, *From Foraging to Agriculture: The Levant at the End of the Ice Age* (1989) (see also Stephen Jay Gould's review of Henry's book in "Down on the Farm," *New York Review of Books* 36 [January 18, 1990]: 26–27). Garret Hardin's "The Tragedy of the Commons," *Science* 162 (December 1968): 1243–48, and Lynn White, Jr.'s "The Historical Roots of Our Ecologic Crisis," *Science* 155 (March 1967): 1203–7, are also worth reading.

Finally, two relevant issues that I think deserve special attention are (a) the role of man-made (so-called cultural) fires in altering and otherwise affecting the biota, both plant and animal, creating in certain regions what amounted to semiagricultural conditions (see, e.g, Lowell John Bean and Harry W. Lawton, "Some Explanations for the Rise of Cultural Complexity in Native California with Comments on Proto-Agriculture and Agriculture," the introduction to Henry T. Lewis, *Patterns of Indian Burning in California: Ecology and Ethnohistory*, Ballena Press Anthropological Papers No. 1, 1973); and (b) the nutritional consequences of the neolithic (for suggestive thoughts, see H. Leon Abrams, Jr., "Vegetarianism: An Anthropological/Nutritional Evaluation," *Journal of Applied Nutrition* 32, no. 2 [1980]:53–87).

~~~~~~~~~~~~

While writing this final essay I had a particularly painful collision with time. I was riding my bicycle down the main street of Princeton in the spirit of the earth when a car ran into me. Pulling on the hand brake, a distraught young woman jumped out and declared that *time* was the culprit. She was late for something-or-other, she exclaimed, and in that blind state had not spotted me. Near tears myself, with an eye on my skinned elbow and feeling a dull ache in my leg, I collected my thoughts. I honestly felt no urge to rebuke her. What I did say probably made her suspect I had hit my head: I said I was a philosopher, and that she mustn't let time tyrannize her. She readily agreed, more from feeling awful, I'm sure, than anything else.

Another Princeton cyclist, well known for his experiments with time, Albert Einstein, once responded—to a newspaper reporter, I think it was—that he often wondered whether the universe was friendly or not. A strange question to ask, a quintessentially humanistic way of seeing things. It is not my question, nor is it the question of this book. I think if my former neighbor had held a different perspective on time, different from the way it has been codified into history, he would not have thought to ask it, either.

I regret being wounded, of course, but I am intrigued that the motorist who hit me blamed the curse of time. At least that's how she saw it, and there lies the key: time as she understood it. I was using my bike to get around town precisely to sense a different sort of time—a small and manifestly hazardous exercise in pacing. Pacing and aesthetics—very important. In my life, these days, I look for the silences in time, where time cannot scratch its name or gain a purchase, like the pauses in the Zuni poet's narrative of the First Time. Or like the silences of mountains. Like all the rest of my kind, I am a born talker, yet my appreciation for the silence during conversation has grown. I am, too, like all humans, a born timekeeper, but in paying attention to the spaces between the ticks and the tocks, training my ear to hear them more than the sounds of time's errands, I believe I have found a more universal time. The silent space between the units of time is a time realm of its own. The units of measurement change, culture to culture, age to age (see G. J. Whitrow, *Time in History: Views of Time from Prehistory to the Present Day*, 1989), but the silent spaces never ever do. That is the secret of meditation, of poetry, and music. This has become for me the *real* time: the silent interstices of timely units, the intervals that defy calculation. It is, moreover, only by perceiving this that I understand the world and my life as an aesthetic, rather than a thing born of *chronos* and *logos*.

History, I have argued, is not an aesthetic nor does it contemplate beauty. Its domain is in the counting, the awesome computational power of the numbers themselves, reminiscent of the time when these seemingly benign units were perceived as gods. Mythology, that timeless thing, is misunderstood: it exists in a different realm from that of the numbers; its domain is in the spaces that no one can quantify. And it is always there. It speaks the universal, and universalizing, language, because it is, above all, poetry and song.

(The Aztec sages knew that well, but were too overwhelmed by the intense gravitational pull of numbered history to alter course.) To live, to think, but most of all to imagine in those incalculable spaces (witness Grey, in Momaday's *Ancient Child*) is to live in an aesthetic: the beauty of the earth. For such is the spirit of the earth.

In all of this, I keep thinking of Niels Bohr's principle of complementarity, which asserts that

> there exist complementary properties of the same object of knowledge, one of which if known will exclude knowledge of the other. We may therefore describe an object like an electron in ways which are mutually exclusive—e.g., as wave or particle—without logical contradiction provided we also realize that the experimental arrangements that determine these descriptions are similarly mutually exclusive. Which experiment—and hence which description one chooses—is purely a matter of human choice. (Heinz R. Pagels, *The Cosmic Code: Quantum Physics as the Language of Nature*, 1982, p. 75)

Bohr's principle holds true for the microcosmic and macrocosmic world of quantum mechanics; does it hold for the contemplation of time on a human scale, I wonder? If not, should that human scale then be redefined, to gear with the microcosmos and macrocosmos? Is that not where mythology makes the most sense, anyway, and history becomes absurd?

This book marks my collision with time and its fretting—history in the largest sense, my life in the smallest—and my effort to learn the meaning of the interstices: to walk in a domain of beauty. When we maintain beauty, for which *Homo sapiens* has an uncannily developed eye and ear and touch (see John E. Pfeiffer, *The Creative Explosion: An Inquiry into the Origins of Art and Religion*, 1982), we live, quite simply, in timelessness. The funny thing is we all know that.

Einstein wondered if the universe is friendly. I would rather ask, as old Albert Dreyer kept needling Eiseley in "The Dance of the

Frogs" (*The Star Thrower*), if the universe dances. Or sings. I am learning, in the quiet spaces between the numbers, that it does. And perhaps I shall, too.

‹‹‹‹‹‹‹‹‹‹‹‹‹‹‹

> With beauty before me, I walk
> With beauty behind me, I walk
> With beauty above me, I walk
> With beauty below me, I walk
> From the East beauty has been restored
> From the South beauty has been restored
> From the West beauty has been restored
> From the North beauty has been restored
> From the zenith in the sky beauty has been restored
> From the nadir of the earth beauty has been restored
> From all around me beauty has been restored.
>
> —traditional Navajo poem

# Notes

〜〜〜〜〜〜

## Chapter 1.
### "An inhuman company of elder things"

Chapter title: Loren Eiseley, *The Star Thrower* (New York: Times Books, 1978), 114.

1. Joseph Epes Brown, "Becoming Part of It," *Parabola* 7 (Summer 1982):12.

2. Marcel Detienne, *The Creation of Mythology* (Chicago: University of Chicago Press, 1986), 9.

3. N. Scott Momaday, *The Ancient Child* (New York: Doubleday, 1989), 121.

4. Reuben Gold Thwaites, ed., *The Jesuit Relations and Allied Documents: Travels and Explorations of the Jesuit Missionaries in New France, 1610–1791*, 73 vols. (Cleveland: Burrows Brothers, 1896–1901), 24:209–11; Frank G. Speck, *Naskapi: The Savage Hunters of the Labrador Peninsula* (Norman: University of Oklahoma Press, 1935), 172; Raymond D. DeMallie, ed., *The Sixth Grandfather: Black Elk's Teachings Given to John G. Neihardt* (Lincoln: University of Nebraska Press, 1984), 92.

5. Robin Ridington, "Fox and Chickadee," in *The American Indian and the Problem of History*, ed. Calvin Martin (New York: Oxford University Press, 1987), 129.

6. Ibid., 130.

7. Ibid., 133.

8. "The War Against the Animals," *Wall Street Journal*, 19 July 1978, p. 22, col. 1.

9. Claude Lévi-Strauss, *Totemism*, trans. Rodney Needham (Boston: Beacon, 1963), 89.

10. Robert Bringhurst, "Sunday Morning," in *Pieces of Map, Pieces of Music* (Toronto: McClelland and Stewart, 1986), 68.

11. Ruth Landes, *Ojibwa Religion and the Midéwiwin* (Madison: University of Wisconsin Press, 1968), 21.

12. Speck, *Naskapi*, 94.

13. James McKenzie, "The King's Posts and Journal of a Canoe Jaunt Through the King's Domains, 1808, the Saguenay and the Labrador Coast . . . ," in *Les Bourgeois de la Compagnie du Nord-Ouest: Recits de Voyages, Lettres et Rapports Inedits Relatifs au Nord-Ouest Canadien*, ed. L. R. Masson (Quebec: De l'Imprimerie Générale A. Coté et Cie, 1889–1890; reprint, New York: Antiquarian Press, 1960), ser. 2, p. 415.

14. Richard K. Nelson, *Hunters of the Northern Ice* (Chicago: University of Chicago Press, 1969), 363.

15. Marshall D. Sahlins, *Stone Age Economics* (Chicago: Aldine/Atherton, 1972), 1–39.

16. Ridington, "Fox and Chickadee," 133–34.

17. Nelson, *Hunters*, 374.

18. Speck, *Naskapi*, 83.

19. Ibid., 76.

20. Stanley Diamond, "Eskimo," in *Totems* (Rhinebeck, N.Y.: Open Book Publications in association with Station Hill Press, 1982), 51.

## Chapter 2.
### *"Creation of a margin"*

*Chapter title:* Frederick Turner, *Beyond Geography: The Western Spirit against the Wilderness* (New York: Viking Press, 1980), 23 (paraphrase).

1. Loren Eiseley, "The Little Treasures," in *Another Kind of Autumn* (New York: Charles Scribner's Sons, 1977), 56.

2. "Natural Magic," interview with David Abram, *Minding the Earth: Newsletter of the Strong Center for Environmental Values*, Berkeley, Calif., vol. 4, no. 2 (June 1983).

3. Ibid.

4. Ibid.

5. Ibid.

6. Loren Eiseley, "The Deer," in *Another Kind of Autumn* (New York: Charles Scribner's Sons, 1977), 77.

7. Lewis Mumford, *The City in History: Its Origins, Its Transformations, and Its Prospects* (New York: Harcourt, Brace and World, 1961), 15–16; Frederick Turner, *Beyond Geography*, 23.

8. John Dunn, *The Oregon Territory, and the British North American Fur Trade; with an Account of the Habits and Customs of the Principal Native Tribes on the*

*Northern Continent* (Philadelphia: G. B. Zieber, 1845), 60–61.

9. Marc Lescarbot, *The History of New France*, 3rd ed., trans. W. L. Grant (Paris: 1609; 3rd ed. 1618; Toronto: The Champlain Society, 1914), vol. 3, p. 214.

10. Ibid., 190.

11. Ibid., 172.

12. Peter Grant, "The Sauteux Indians (about 1804)," in *Les Bourgeois de la Compagnie du Nord-Ouest: Recits de Voyages, Lettres et Rapports Inedits Relatifs au Nord-Ouest Canadien*, ed. L. R. Masson (Quebec: De l'Imprimerie Générale A. Coté et Cie, 1889–1890; reprint, New York: Antiquarian Press, 1960), ser. 2, p. 331.

13. James McKenzie, "The King's Posts and Journal of a Canoe Jaunt," ser. 2, p. 425.

## Chapter 3.
### "Time was the god"

*Chapter title:* Loren Eiseley, "The Maya," in *Another Kind of Autumn* (New York: Charles Scribner's Sons, 1977), 23.

1. Matthew James Dennis, "Cultivating a Landscape of Peace: The Iroquois New World, " (Ph.D. diss., University of California at Berkeley, 1986), 29–33.

2. Eiseley, *The Star Thrower*, 218, 179.

3. Marshall Sahlins, *Islands of History* (Chicago: University of Chicago Press, 1985), 80, 109, 49, 50.

4. Eiseley, *The Star Thrower*, 178.

5. Eiseley, "The Maya," 23-24.

6. Eliot Weinberger, trans., "Han Yu's Address to the Crocodiles," in *Works on Paper, 1980–1986* (New York: New Directions, 1986), 41–42. Italicized paragraphs are translator's words.

## Chapter 4.
### "I am inside you, have no fear"

1. John M. Coetzee, "The Narrative of Jacobus Coetzee," in *Dusklands* (New York: Penguin Books, 1974), 51–125.

2. Ibid., 78–80.

3. Francis Lee Utley, Lynn Z. Bloom, and Arthur F. Kinney, eds., *Bear, Man, and God: Eight Approaches to William Faulkner's "The Bear,"* 2nd ed. (New York: Random House, 1971), 14.

4. Eiseley, *The Star Thrower*, 63.

5. Utley, Bloom, and Kinney, *Bear, Man, and God*, 15.

6. Leslie E. Gerber and Margaret McFadden, *Loren Eiseley* (New York: Frederick Ungar, 1983), 157.

7. Thomas W. Overholt and J. Baird Callicott, *Clothed-in-Fur and Other Tales: An Introduction to an Ojibwa World View* (Washington, D.C.: University Press of America, 1982), 74.

8. Henry David Thoreau, *Walden; or, Life in the Woods* (New York: Holt, Rinehart and Winston, 1948), 42; Utley, Bloom, and Kinney, *Bear, Man, and God*, 81, 39.

9. A. Irving Hallowell, "Bear Ceremonialism in the Northern Hemisphere," *American Anthropologist*, n.s., vol. 28, no. 1 (1926):1–175.

10. Gary Snyder, *Good, Wild, Sacred* (Madley, Hereford, England: Five Seasons, 1984), no pagination. See the same story in slightly reworded and more accessible form in Snyder, *The Practice of the Wild* (San Francisco: North Point, 1990), 82–83.

11. Ruth Murray Underhill, *Singing for Power: The Song Magic of the Papago Indians of Southern Arizona* (Berkeley and Los Angeles: University of California Press, 1938), 6–7.

12. Snyder, *Good, Wild, Sacred*, no pagination; Gary Snyder, *The Old Ways: Six Essays* (San Francisco: City Lights Books, 1977), 35, 64.

13. N. Scott Momaday, *The Way to Rainy Mountain* (New York: Ballantine Books, 1969), 2, 42.

14. Ibid., 113.

15. Robert Bringhurst, "Dull Myths and Sharp Myths" (March 1988, mimeographed), 2-3. This essay was published as "Myths Create a World of Meaning," *Globe and Mail* (Toronto), 7 May 1988.

16. Detienne, *The Creation of Mythology*, 114–15.

17. Thoreau, *Walden*, 13.

18. Lewis Hanke, ed., *History of Latin American Civilization: Sources and Interpretations*, vol. 1, *The Colonial Experience* (Boston: Little, Brown, 1967), 123–25.

19. Randall Jarrell, "Bats," in *The Complete Poems* (New York: Farrar, Straus and Giroux, 1969), 314.

20. Robert Bringhurst, "Thirty Words," in *Pieces of Map, Pieces of Music* (Toronto: McClelland and Stewart, 1986), 66.

## Chapter 5.
### *"We have come to where history ends"*

*Chapter title*: Loren Eiseley, "Druid Born," in *Another Kind of Autumn* (New York: Charles Scribner's Sons, 1976), 64.

1. Robert Bringhurst, "A Quadratic Equation," in *The Beauty of the Weapons: Selected Poems, 1972–82* (Toronto: McClelland and Stewart, 1982), 22.

2. Loren Eiseley, "The Eye Detached," in *Another Kind of Autumn* (New York: Charles Scribner's Sons, 1977), 58; Loren Eiseley, *The Immense Journey* (New York: Random House, 1957), 59.

3. Eiseley, *The Star Thrower*, 175.

4. Ibid., 176, 226.

5. Thoreau, *Walden*, 113; Robert Bringhurst, "Xenophanes," in *The Beauty of the Weapons: Selected Poems, 1972–82* (Toronto: McClelland and Stewart, 1982), 65.

6. Bringhurst, "Xenophanes," 65.

7. Loren Eiseley, "Men Have Their Times," in *Another Kind of Autumn* (New York: Charles Scribner's Sons, 1977), 33.

8. Eiseley, *The Star Thrower*, 237–38.

9. Ibid., 43.

10. George Santayana, *The Life of Reason, or the Phases of Human Progress: Introduction and Reason in Common Sense*, 2nd ed. (New York: Charles Scribner's Sons, 1924), 284.

11. Ralph Waldo Emerson, "History," in *The Portable Emerson*, ed. Mark Van Doren (New York: Penguin Books, 1977), 163–64.

12. Robin Fox, *The Search for Society: Quest for a Biosocial Science and Morality* (New Brunswick: Rutgers University Press, 1989), 232.

13. Ibid., 232, 219.

14. Ibid., 212–41.

15. Thoreau, *Walden*, 272; Eiseley, *The Immense Journey*, 21; Eiseley, "Druid Born," 64.

16. Eiseley, *The Star Thrower*, 238.

Calvin Luther Martin is associate professor of history at Rutgers University, New Brunswick, N.J. He is the author of *Keepers of the Game: Indian-Animal Relationships and the Fur Trade* (1978, winner of the American Historical Association's Albert J. Beveridge award) and the editor of *The American Indian and the Problem of History* (1987). He has been awarded fellowships from the National Endowment for the Humanities, the John Simon Guggenheim Memorial Foundation, the American Council of Learned Societies, and the Henry E. Huntington and Newberry libraries.

*In the Spirit of the Earth*

Designed by Ann Walston

Composed by Brushwood Graphics, Inc.,
in Berkeley Old Style

Printed by the Maple Press Company
on 55-lb. S.D. Warren's Antique Cream Sebago
and bound in Holliston Aqualite and G.S.B. cloth
with James River Papan endsheets